INTRODUCTION TO HEAT TRANSFER

Introduction to
HEAT TRANSFER

AUBREY I. BROWN
Professor of Mechanical Engineering
The Ohio State University

SALVATORE M. MARCO
Professor of Mechanical Engineering
Chairman, Department of Mechanical Engineering
The Ohio State University

INTERNATIONAL STUDENT EDITION

McGRAW-HILL BOOK COMPANY, INC.

New York   Toronto   London

KŌGAKUSHA COMPANY, LTD.

Tokyo

INTRODUCTION TO HEAT TRANSFER

INTERNATIONAL STUDENT EDITION

Exclusive rights by Kōgakusha Co., Ltd., for manufacture and export from Japan. This book cannot be re-exported from the country to which it is consigned by Kōgakusha Co., Ltd.

TOSHO   INSATSU
PRINTING CO., LTD.
TOKYO   JAPAN

PREFACE TO THE THIRD EDITION

The authors' purpose, as in the preparation of the earlier editions of this book, has been to present the fundamentals of heat transfer in a manner readily understandable to engineering students and engineers in practice who have completed the courses in physics and mathematics usually required in the first two years of engineering curricula and an elementary course in heat power.

In this book the emphasis is placed upon acquiring a clear conception of the manner in which heat is transmitted and upon the development of the fundamental mathematical expressions that apply to the calculations of heat transfer through clean surfaces. The attempt has been made to supply data from the most authoritative sources. The reader is warned, however, that correlation of the results of different experimenters' work is in many instances very poor and that calculated results are subject to error from this source. He should also be aware that errors are sure to be present both in experimental and in calculated results wherever such uncertain factors as the condition of surfaces have not been taken into proper account. Little space has been devoted to the correlation of data from various sources and to the effects of uncommon and variable influences upon heat transfer which may at times be present, such material being considered to fit more appropriately into a more advanced text. It is the opinion of the authors that the engineer who engages in work in heat transfer learns soon enough of the difficulties encountered in his attempt to find close agreement between calculated and experimentally determined results, whether through variations in technique, lack of accurate reference data, or the oversight of unusual factors that may be present; but the starting point for successfully overcoming these difficulties lies in thorough knowledge of the ways by which heat is transferred and in an acquaintance with the fundamental mathematical expressions.

Most of the material here presented has been used with satisfactory results in a required course for third-year mechanical-engineering students at The Ohio State University, following a course in fluid mechanics and several courses in heat power.

The authors have included numerous illustrative examples, which are presented with considerable detail. Special emphasis has been placed upon the use of a consistent system of units. Although data are included on the physical properties of many of the more common materials, no attempt has been made to provide complete reference data, and the reader is referred to the engineering handbooks or to the International Critical Tables for more extensive data.

Solicitation of the opinions of users of the former editions has revealed that although a few suggest a more advanced or rigorous treatment of the subject by earlier introduction and greater application of the unsteady-state considerations, the majority opinion definitely favors keeping the treatment at the same level as in the former editions. The authors have therefore endeavored to adhere to this opinion with the hope that for many students this introduction to the subject will serve as a foundation for more advanced study.

In the preparation of this edition the authors have attempted to expand and clarify portions which in the opinion of a number of users have needed additional explanation or revision. Two new chapters have been added; notable additions have been made to two other chapters; and many articles have been rearranged.

A new chapter on fluid flow in the convection process is intended to serve as a review of the principles of fluid flow that are pertinent to a study of heat transfer. Otherwise, it provides introductory material in the field of fluid mechanics for those students who have had no course in that subject.

The second new chapter deals with graphical and numerical methods applied to problems of heat conduction. The chapter which formerly dealt mainly with dimensional analysis has been amplified by additional material on fundamental units and their relationships; and in the chapter on the application of the principles of heat transfer to design problems a section has been added dealing with the design of electrical transformers.

Throughout the book, revisions in symbols and abbreviations

have been made in accordance with the recommended practices of the American Standards Association.

Any claim of originality in this text must lie primarily in the method of presentation of the subject. The authors have drawn freely from the publications of the national technical societies and from many other sources. They have made frequent references to the sources of information but wish to acknowledge their special indebtedness to the following writers who have contributed so largely to the literature on heat transfer, namely, W. H. McAdams and H. C. Hottel of the Massachusetts Institute of Technology and the late Max Jakob of the Illinois Institute of Technology. The authors wish also to thank those users who have given their much appreciated criticisms of the text and their valued suggestions for its improvement. Special thanks are due to Professor E. B. Penrod of the University of Kentucky and to our colleagues at The Ohio State University, Professors R. H. Zimmerman, M. L. Smith, G. D. Hudelson, and L. S. Han.

Aubrey I. Brown
Salvatore M. Marco

CONTENTS

NOMENCLATURE

Units of Measurement.—Much of the literature on the subject of heat transfer has had its origin in Germany or in the United States, two countries in which different systems of measurement are ordinarily employed. The language of the physicists, chemists, and engineers who have been the chief contributors to this literature also has often been at variance. As a result, considerable confusion has at times developed, particularly in the minds of engineering students who are confronted with such a variety of terms and units with which they have had little acquaintance.

This text has been written primarily for the use of engineering students and engineers. The authors have endeavored, therefore, to express all data derived from whatever source in engineering terms and in a consistent system of units. Heat quantities are here expressed in British thermal units (Btu); temperatures are in degrees Fahrenheit; linear dimensions are in feet; weights are in pounds; and the unit of time is the hour. Some inconvenience may be encountered because of the expression of velocities in feet per *hour* instead of feet per second, as is customary in recording velocities of liquids, or in feet per minute, as is most common in the measurement of the flow of gases; but consistency is felt to be more desirable here than convenience. Since the hour is the common basis for the measurement of heat flow, it will, except in rare, noted instances, be used as the standard unit of time.

Confusion sometimes arises because of the multiplicity of terms used in the various fields of science to denote mass, force, and weight. In this text the standard unit of measurement of all three properties is the pound. Distinction is made, however, between the pound force and the pound mass. Mass is a measure of the amount of matter, fixed, in this country, by act of

Congress by reference to a standard pound body carefully preserved at the U.S. Bureau of Standards.   Force also is a measure of a fixed value, the pound force being defined as the force required to support the standard pound body against gravity, *in vacuo*, in the standard locality.   A pound weight may or may not represent a fixed amount of matter, depending upon the type of weighing device employed; for a platform balance gives the same indication of weight in all localities, whereas a spring balance is influenced by a variable force of gravity in different locations. It should be observed that mass and weight, when expressed in pounds, have the same numerical value, provided the object when being weighed is balanced against standard weights as is customarily the case when objects are weighed on platform scales.

The symbols and units applied to the various quantities, used in this text, are shown in the following table, Table A.

TABLE A. HEAT-TRANSFER NOMENCLATURE

(Greek symbols are grouped together following the English)

Symbol	Quantity	Units
a	Absorptivity	
a	Modulus $g\beta\rho^2 c_p/\mu k$	
a	Coefficient	
A	Area of heat-transfer surface or of cross section	sq ft
b	Coefficient (Chap. 9)	
B	Coefficient (Chap. 7)	
c	Constant (Chap. 9)	
c_p	Specific heat at constant pressure	Btu/(lb$_m$)(°F)
c_v	Specific heat at constant volume	Btu/(lb$_m$)(°F)
C	Thermal capacitance (Chap. 14)	Btu/°F
C	Conductance of air space	Btu/(hr)(sq ft)(°F)
C	Coefficient	
C_t	Thermal conductance (Chap. 3)	Btu/(hr)(°F)
d	Differential	
d	Thickness (Chap. 8)	ft
d	Coefficient (Chap. 9)	
D	Diameter	ft
e	Length of an edge (Chap. 3)	ft
e	Base of natural log (2.718)	
E	Total emissive power	Btu/(hr)(sq ft)
f	Friction factor	
f	Function of	

TABLE A. HEAT-TRANSFER NOMENCLATURE. *(Continued)*

Symbol	Quantity	Units
F	Degrees Fahrenheit	°F
F	Correction factor (Chap. 11)	
F_A	Configuration factor	
F_e	Emissivity factor	
g	Acceleration of gravity	ft/sec² or ft/hr²
g_c	Dimensional constant (32.1739)	$(lb_m/lb_f)(ft/sec^2)$ or $(lb_m/lb_f)(ft/hr^2)$
h	Surface coefficient	Btu/(hr)(sq ft)(°F)
h_c	Surface coefficient of convection	Btu/(hr)(sq ft)(°F)
h_r	Surface coefficient of radiation	Btu/(hr)(sq ft)(°F)
h	Head (Chap. 6)	ft of fluid
I	Radiation intensity	Btu/(hr)(sq ft)
k	Thermal conductivity	Btu/(hr)(sq ft)(°F/ft)
K	Thermal conductance (Chap. 14)	Btu/(hr)(°F)
l	Length	ft
L	Length	ft
L	Heat loss (Chap. 12)	watts/sq in.
m	Exponent (Chap. 7)	
M	Mass	lb_m
M	Coefficient (Chap. 14)	
n	Number	
n	Exponent	
N	Number	
N_{Gr}	Grashof number $g\beta\theta L^3\rho^2/\mu^2$	
N_{Nu}	Nusselt number $h_c D/k$	
N_{Pr}	Prandtl number $c_p\mu/k$	
N_{Re}	Reynolds number $DV\rho/\mu$	
N_{St}	Stanton number $h_c/Vc_p\rho$	
p	Pressure	lb_f/sq ft
P	Perimeter	ft or in.
P	Temperature-difference ratio (Chap. 11)	
q	Time rate of heat transfer	Btu/hr
Q	Volume flow rate	cu ft/hr
Q_g	Heat generated	Btu/hr
r	Radius	ft or in.
r	Reflectivity	
r	Latent heat of vaporization (Chap. 10)	Btu/lb_m
R	Ratio of outside to inside surface (Chap. 12)	
R	Residual temperature (Chap. 14)	°F
R_t	Thermal resistance	(°F)(hr)/Btu
S	Surface	sq ft
t	Time	hr

TABLE A. HEAT-TRANSFER NOMENCLATURE. (*Continued*)

Symbol	Quantity	Units
t	Temperature	°F
T	Absolute temperature ($t + 460$)	°F abs
U	Transmittance (over-all coefficient)	Btu/(hr)(sq ft)(°F)
v	Volume per lb mass	cu ft/lb_m
V	Velocity	ft/hr or ft/sec
w	Weight flow rate	lb_f/hr
W	Weight	lb_f
W	Rate of heat transfer (Chap. 12)	watts/sq in.
x	Thickness	ft
x,y,z	Rectangular coordinates	
z	Elevation	ft
α	A constant	
α	Thermal diffusivity	sq ft/hr
α	An angle	
β	Coefficient of expansion	1/°F abs (for perfect gases)
β	An angle	
β	Factor (Chap. 7)	
γ	Ratio of specific heats c_p/c_v	
γ	Specific weight	lb_f/cu ft
Δ	Difference	
Δt	Temperature difference	
ϵ	Emissivity	
ϵ	Surface roughness	ft
θ	Temperature	°F
θ	An angle	
λ	Wave length	microns
μ	Absolute viscosity	lb_m/(ft)(hr)
μ	Microns, 0.0001 cm	
ν	Frequency	1/sec
π	3.14159	
ρ	Density	lb_m/cu ft
Σ	Summation of	
σ	Stefan-Boltzmann constant, 0.173×10^{-8}	Btu/(hr)(sq ft)(T^4)
ϕ	Function of	
ϕ	An angle	
ψ	Function of	
ψ	An angle	

CHAPTER 1

MODES OF HEAT TRANSFER

1-1. General. When energy is transferred from one body to another by virtue of a temperature difference existing between them, it is said that *heat* is transferred. The fundamental law governing this phenomenon is that heat may be transferred from a high-temperature region to one of lower temperature but never from a low-temperature region to one of higher temperature. Since the transfer of heat is in reality a transfer of energy, it necessarily follows the law of conservation that the heat emitted by the high-temperature region must be exactly equal to the heat absorbed by the low-temperature region. A study of heat transmission must therefore be built upon these two fundamental axioms. The science of heat transfer concerns itself primarily with the determination of the quantity of heat transmitted from one region to another.

The transfer of heat from one region to another occurs in many commonplace processes; and the principles of heat transfer are involved in the design of many forms of industrial and commercial equipment. The heat for boiling water in a steam boiler must be transmitted from the fire and hot gases, through tubes and plates, and into the water. A similar transfer takes place in closed feed-water heaters and in economizers; in surface condensers for steam, ammonia, and other refrigerants; in steam and hot-water heating systems; through the walls of buildings, outward when heated in winter and inward when cooled in summer; in evaporators or stills; in the cooling of compressor and engine cylinders; in the lubrication of bearings; and in many other pieces of apparatus.

In most of the apparatus named above, some solid separates the high-temperature "source" and the lower temperature "receiver." In those cases where it is desired to transmit heat,

1

separating solids are used that transmit heat readily. However, in those cases where it is desired to prevent the transmission of heat, an attempt is made to use separating solids that do not transmit heat easily. The walls of buildings, for example, are "heat-insulated" to minimize the outward flow of heat in the winter and the inward flow in the summer.

The flow of heat may occur in one or more of three essentially different ways. These three methods of heat transmission are **conduction, radiation,** and **convection.** Although for the sake of convenience the field of heat transmission is considered under these three divisions, actually nature does not recognize this attempt at simplification. Conduction problems may involve radiation and convection, and convection problems invariably involve conduction and radiation. In some cases it is possible to determine as separate items the heat transmitted by each of the different methods; in many other cases it is possible to determine only the over-all heat transmission. In any case the solution of heat-transfer problems at the present time involves the application of laws governing conduction, radiation, and convection that have been established for many years, together with empirical data that have been determined accurately only in recent years. There are still a great many unknowns in the field of heat transfer. Many investigators are working at the present time to determine some of these unknown factors.

1-2. Conduction, Radiation, and Convection. To facilitate the understanding of heat-transfer problems, the theory of each process will be treated separately, and then its application to the solution of actual problems will be considered, together with the experimental work of various investigators. This separation into three divisions is possible because each process involves a different method in transferring heat from one place to another.

1-3. Conduction. Heat may flow from a high-temperature region to a lower-temperature region within a body by conduction. This process is not a simple one, and its exact nature, which is dependent on the type of material, is not completely understood. In general, the particles of matter (molecules, atoms, and electrons) in the high-temperature region, being at higher energy levels, will transmit some of their energy to the adjacent lower-temperature regions. In the case of conduction in gases the interchange of kinetic energy by molecular collision is probably the predominant mechanism. In nonmetallic liquids the process involves the propagation of lattice-vibration waves as well as molecular collision. In nonmetallic solids the primary mechanism is probably lattice-vibration-wave propagation, while in metallic solids the flow of free electrons is primarily responsible for heat conduction.

Heat may also be transferred by conduction from one body at higher temperature to another at lower temperature if the two bodies are in physical contact. In this process the heat flows by conduction through the boundaries of the two bodies.

The laws governing internal conduction are well established and can be derived through mathematical considerations, but the laws governing conduction through boundaries are not so well understood.

1-4. Radiation. Heat may pass from a higher-temperature body, through space, to other bodies at lower temperatures at some distance away without necessarily warming the medium within the space. This phenomenon is called radiation, and it operates by virtue of a wave motion in a manner very similar to light radiation. Heat radiation may take place through a vacuum, through some gases, and through a few liquids. The most common example of this form of heat transmission is the passage of the sun's heat to the earth. The methods of computation for heat radiation are somewhat similar to those utilized for light radiation.

1-5. Convection. Heat may be carried from a higher- to a lower-temperature region within a gas or a liquid by moving masses of the fluid. The water heated at the bottom of a vessel moves to the top because its density is less than that of the colder water above it. This process is called natural convection when circulation is caused merely by differences in density which accompany changes in temperature. If circulation is caused by fans, blowers, pumps, etc., the heat is then said to be transferred by forced convection.

Since heat transfer by convection depends upon the motion of fluids, it is necessary to understand some of the laws of fluid flow before the laws governing heat transfer by convection can be understood.

1-6. Evaporation and Condensation. Heat transfer resulting from the evaporation of liquids and from the condensation of vapors should be differentiated from the simpler forms of free and forced convection. Since evaporation and condensation necessarily involve the flow of fluids, they will be considered special forms of convection, although by some writers they are classed as two separate, additional modes of heat transfer.

1-7. The Necessity for Separate Determinations of Conduction, Radiation, and Convection. In any study of heat transmission a knowledge of the laws of the three modes of transmission is essential. It has already been indicated that when heat is transferred from a higher- to a lower-temperature region, more than one of the three modes of transmission is usually involved. The importance of determining separately the heat transferred by conduction, by radiation, and by convection, however, may not be apparent. The reason for calculating the heat transfer by the three modes separately is that even with a fixed difference in temperature of the two regions, different amounts of heat may be transferred when variation is made in any factor that affects any one of the three modes independently of the others.

The two following examples will illustrate the manner in which a change in the thickness of a conductor of heat, affecting conduction only, may influence the over-all transmission.

It is common knowledge that copper is a good conductor of heat but a poor radiator. If heat is to be transmitted from warm air on one side of a sheet of copper to cooler air on the other, what will be the effect of changes in the thickness of the copper? In this case the copper sets up a relatively insignificant resistance to the passage of heat by conduction; the chief barriers occur at the surface of the metal because of the poor radiating value of copper and because air (unless at a high velocity) is not a very effective medium for conveying heat by convection. In the light of these observations it is apparent that the thickness of the metal has little bearing upon the transmission of heat; in fact, within the accuracy of slide-rule calculations no more heat will be transmitted through a copper sheet 0.002 in. thick than through a copper plate 2 in. thick, for identical conditions of the air on the two sides.

In contrast with the foregoing example, if a substance of low thermal conductivity, such as corkboard, is substituted for the copper, the resistances to heat transmission at the surfaces, by radiation and convection, are small in comparison with the barrier to heat flow by conduction which is set up by the material of low conductivity. In consequence, in a given time and for a given difference in the temperatures of the air on the two sides, the amount of heat transferred through a 4-in. thickness of corkboard will be practically one-half of that which will pass through corkboard 2 in. thick.

The fundamental value usually sought in problems of heat transmission is the over-all coefficient U, which is the amount of heat transmitted per unit area in unit time for a difference of 1°F in the temperatures of the hotter and cooler mediums. For a specific set of conditions U may be determined by test, but for any other set of conditions it can be predicted accurately only by application of knowledge of the separate effects of conduction, radiation, and convection.

In most instances the transfer of heat takes place between fluids that are separated by a layer of solid material. Heat flows from the hotter fluid to the surface of the solid by the process of convection or by both convection and radiation. The heat then flows through the solid material by conduction and is finally delivered from the surface to the cooler fluid by convection or by both convection and radiation.

The exchange of heat between the fluid and the surface of the solid is called **surface conductance.** Where only one mode of heat transfer is involved, surface conductance may be noted specifically as **surface conductance due to convection** or as **surface conductance due to radiation.** Although the term *surface conductance* is generally applied without regard to the process involved, it is appropriately used only where convection alone occurs, for convection is in reality a process of conduction of heat through a film of fluid that adheres to the surface of the solid. As applied to radiation, the term *surface conductance* is a misnomer, since it implies a process that is entirely absent.

The surface conductance when expressed in Btu per hour per square foot of surface for a difference of 1°F in the temperatures of the fluid and the adjacent surface is known as the **surface coefficient,** or **film coefficient,** and is denoted by the symbol h. The surface coefficient of convection only is h_c; and h_r is the surface coefficient of radiation.

The **coefficient of thermal conductivity** k, for the solid material, represents the amount of heat transmitted by conduction, in Btu per hour, from 1 sq ft of surface on one side of the solid to an equal extent of surface on the other side of the solid 1 ft distant, for a temperature change of 1°F. If the thickness of the solid is x ft, then k/x represents the heat transmitted in Btu/ (hr)(sq ft)(°F).

Figure 1-1 shows the manner in which the temperature changes along the path of heat flow between two fluids at temperatures t_1 and t_4 when separated by a solid layer of thickness x ft.

A drop in temperature from t_1 to t_2 must occur between the region at temperature t_1 and the adjacent surface at temperature t_2 in order to induce the flow of heat through the surface resistance by convection or by radiation or by both processes. A

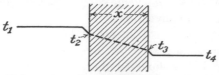

FIG. 1-1. Temperature changes along the path of heat flow.

further decrease in temperature must be present between the surfaces at temperatures t_2 and t_3 if heat is to flow by conduction through the solid material.   Finally, the temperature t_4 must be lower than the surface temperature t_3 in proportion to the resistance to flow of heat encountered at the cooler of the two surfaces of the solid.

In keeping with the law of conservation of energy, for conditions of steady flow, the same amount of heat must flow during any given time through each of the three barriers pictured by the three temperature changes. If the rate of heat flow through unit area of the surface is represented by q/A, the flow through the three successive barriers may be stated as

$$\frac{q}{A} = h_1(t_1 - t_2)$$

$$= \frac{k}{x}(t_2 - t_3)$$

and $$= h_2(t_3 - t_4)$$

where h_1 and h_2 = surface coefficients for the two surfaces.

k = coefficient of thermal conductivity.

x = thickness of the solid.

The rate of heat flow may be expressed also as

$$\frac{q}{A} = U(t_1 - t_4)$$

where U = over-all coefficient of heat transmission.

Where three of the four terms of each of the foregoing expressions are known—considering k/x and q/A as single terms—the fourth term is established. These four simple expressions, therefore, may be applied to determine an unknown temperature, the surface coefficient, the coefficient of thermal conductivity, the over-all coefficient of transmission, or the rate of heat flow.

The following chapters are devoted to methods of determining the values of the surface coefficients, the thermal conductance through solid barriers, the over-all coefficients of transmission, and the rate of heat flow.

CHAPTER 2

HEAT TRANSFER BY CONDUCTION

2-1. General. When heat flows into one surface of a body and at the same time flows out of another surface of the body, it encounters three separate resistances. First of all there is a resistance to the inflow of heat at the hot surface. Then the body itself, which acts as a conveyor, offers a resistance to the flow of heat from the hot surface to the cold surface. Finally, there is a resistance to the outflow of heat at the cold surface. The problem of determining the heat flow within the body will be considered first, and the flow through the surfaces will be considered later.

2-2. General Equations for Conduction. The flow of heat between two points in a body of homogeneous material is analogous to the flow of electricity in a conductor. In each case the quantity flowing per unit of time is proportional to the following quantities:

1. The conductivity of the material (in the case of electricity the electric conductivity and in the case of heat flow the thermal conductivity).

2. The area of the conductor perpendicular to the path of the flow.

3. The voltage gradient in the case of electric flow and the temperature gradient in the case of heat flow.

If the amount of heat flowing into a body is exactly equal to the heat flowing out, then the temperature will be different at different points within the body but for any given point the temperature will remain constant. This condition is called the *steady state* and refers only to those cases where the temperature at any given point within a body is independent of the time. The quantity of heat, above any fixed datum, in such a body does not vary with time but remains constant. If, on the other hand,

8

the inflow and outflow of heat are not equal, the temperature from point to point, the temperature at any given point, and the heat content of the body all vary with the time, then heat is said to be flowing in the *unsteady state*.

All practical solutions of heat-conduction problems are special solutions of the unsteady-state equation. This basic equation, which was first presented by Fourier,[1]* will be developed here for the special cases of steady state first, and the case of varying heat flow will be taken up later.

As already stated, the quantity of heat that will flow by conduction per unit of time is proportional to the thermal conductivity of the material, the area of the conductor normal to the path of flow, and the temperature gradient at the area being considered. It is well at this point to define each of these variables.

The coefficient of thermal conductivity is defined as the quantity of heat that will flow across unit area in unit time if the temperature gradient across this area is unity. In this text the units of this coefficient will always be Btu per hour per square foot per degree Fahrenheit per foot and will be designated by the letter k. It will be shown later that the numerical value of k is not constant but depends upon a number of factors, one of which is the material and another is the temperature.

The area considered is always the area normal to the direction of flow. This area, designated by the letter A, is measured in square feet.

The temperature gradient may be defined as the rate of change of temperature with change in distance, or more specifically the infinitesimal change in temperature dt in passing through the infinitesimal distance dL. The temperature gradient is designated by the symbol dt/dL and is always measured in the direction of flow. For the purpose of this text the units of dt/dL will always be degrees Fahrenheit per foot.

The rate of heat transmission by conduction across an area A (square feet) of any homogeneous material is given by Fourier's law as

$$q = -kA \frac{dt}{dL} \quad \text{Btu/hr} \quad (2\text{-}1)$$

Since the temperature gradient is measured in the direction of

* Superior numbers correspond to the bibliography cited at the end of each chapter.

flow and since dt, the temperature change, is negative, i.e., the temperature decreases in passing through the positive distance dL, it is necessary to introduce the negative sign in order that q, the quantity of heat flowing, may be positive.

Equation (2-1) is the fundamental equation for all heat-conduction problems. The solution of general problems is not possible with Eq. (2-1) as it stands because it requires the measurement of dt/dL at all points on the area A as well as the determination of the value of k for the particular conditions of the problem. There are, however, several special cases of heat conduction, very commonly met in practice, that can be solved by proper transformation of and substitution in Eq. (2-1). It is necessary, however, to understand what k is, how it is measured, and the limitations of its measurement before Eq. (2-1) can be used intelligently.

2-3. Coefficient of Thermal Conductivity k and Its Measurement. The coefficient of thermal conductivity has been defined as the quantity of heat that will flow across unit area in unit time if the temperature gradient between the two surfaces through which heat is flowing is unity. Its numerical value is different for different materials and varies somewhat for the same material at different temperatures. Numerical values of k have been determined experimentally, by various investigators, by one of several methods,[2] depending on the material being investigated. If the material is a metal, k is determined by electrically heating one end of a bar of the metal and cooling the other end with a stream of water. The surface of the bar is insulated, and the heat lost through the insulation is accounted for. The rate of heat flow is measured, and the temperatures of two points along the bar, a known distance apart, are determined. Equation (2-1) is then used to calculate the average thermal conductivity for the given temperature range. Usually the difference between the two temperatures is so small that k is practically the thermal conductivity at the mean temperature indicated by the arithmetical average of the two temperatures measured.

When a nonmetallic solid is being investigated, the coefficient of thermal conductivity is determined usually by the guarded-hot-plate[3] method. The test specimen is made into a thin plate and placed between two plates, one of which is heated electrically while the other is cooled with water. The heating element is made up of a square plate A (Fig. 2-1) that is completely surrounded by an independently heated plate B in the form of a square guard ring. The temperature of the two sides of the test plate, the rate of heat input to the inner hot plate, the area exposed to the inner plate, and the thickness of the specimen are used to determine the coefficient of thermal conductivity by substitution in Eq. (2-1). The purpose of the outer heated ring is to maintain the outer portion of the test specimen at such a temperature that there will be no flow of heat within the test specimen adjacent to the inner ring other than in a direction perpendicular to the surface. The arrangement shown in Fig. 2-1, whereby two similar test specimens are tested simultane-

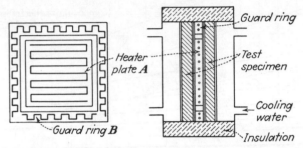

FIG. 2-1. Guarded hot plate.

ously, provides additional assurance that all the heat output of the heater plate will pass through the test specimens.

For measuring the thermal conductivity of gases and liquids the method consists of stretching a fine wire along the axis of a metal tube of small bore. The tube is horizontal and is filled with the fluid being tested. The determination of k then consists of measuring the electrical input to the wire necessary to maintain a measured temperature difference between the wire and the tube wall. Proper precautions are taken to prevent heat flow through the ends of the tube by a device similar to that used in the guarded-hot-plate method. The proper values are substituted in Eq. (2-1), and it is solved for k. As in the previous cases, k is the average thermal conductivity within the given temperature range, and for small temperature differences it is practically equal to the thermal conductivity at the mean of the two temperatures measured.

2-4. Thermal Conductivity of Solids. While there are several methods for determining the thermal conductivities of materials, most of the values of k shown in the tables that follow have been determined by the methods described in Art. 2-3. Some experimentally determined values of k for solids are given in Table 2-1 and Figs. 2-2 and 2-3. An inspection of these values shows that for solids k varies from about 0.014 Btu/(hr)(sq ft)(°F/ft) for silica aerogel, the poorest conductor (best insulator), to 226 Btu/(hr)(sq ft)(°F/ft) for pure copper and 244 Btu/(hr)(sq ft) (°F/ft) for pure silver, the best conductors known. The solid

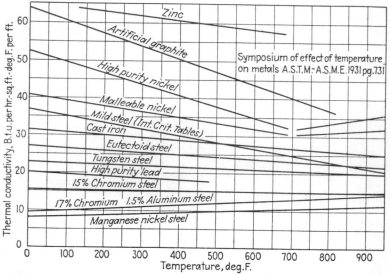

FIG. 2-2. Change in thermal conductivity with change in temperature.

materials included in Table 2-1 are divided into three groups, namely, metals and alloys, which have the highest thermal conductivities; structural and heat-resistant substances, which have intermediate thermal conductivities; and insulating materials, which have the lowest thermal conductivities.

In general, the thermal conductivities of pure metals are high at low temperature and decrease as the temperature increases, while for alloys the reverse is usually true. There are some exceptions to this generalization, as may be seen in Figs. 2-2 and 2-3. Experiments indicate that minute amounts of impurities in the pure metals will very greatly decrease the thermal conductivity. For example, small traces of arsenic in copper will reduce the thermal conductivity to approximately one-third of

that for pure copper. A comparison of the curves for pure
copper and 90-10 brass in Fig. 2-3 shows that the addition of 10
per cent of zinc, which has a thermal conductivity of the order
of 65 at 50°F, to pure copper, which has a thermal conductivity
of 225, results in a brass having a thermal conductivity of the
order of 60 Btu/(hr)(sq ft)(°F/ft). Thus it is seen that the
thermal conductivity of an alloy is not equal to the sum of the

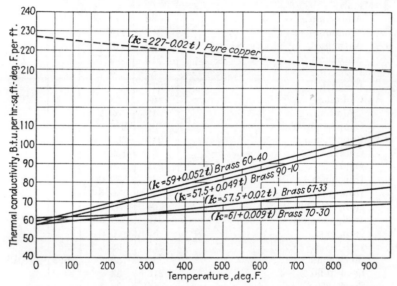

Fig. 2-3. Change in thermal conductivity with change in temperature
(copper and brass).

thermal conductivities of each constituent multiplied by the
percentage of that constituent in the alloy. There is definite
experimental evidence that, for metals, the thermal conductivity
and the effect of temperature on the thermal conductivity are
influenced very greatly by changes in the crystal structure. This
is undoubtedly the reason for the sharp change in slope of the
thermal-conductivity curve for nickel at approximately 700°F
shown in Fig. 2-2. It is to be expected, therefore, that the thermal conductivity of any particular metal or alloy will depend
upon the type of heat-treatment and the type of stress history to
which it has been subjected, as well as the temperature.

At the present time there is no theoretical concept by means
of which the effect of temperature on the thermal conductivity

TABLE 2-1. THERMAL CONDUCTIVITIES OF SOLIDS*

$k = $ Btu/(hr)(sq ft)(°F/ft)

Substance	Temperature, °F	k	k_0	α
a. Metals:				
Aluminum....................	32	117		
Aluminum....................	212	119		
Antimony....................	32	10.6		
Antimony....................	212	9.7		
Bismuth....................	64	4.7		
Bismuth....................	212	3.9		
Cadmium....................	64	53.7		
Cadmium....................	212	52.2		
Copper, pure....................	32	226	227	−0.02
Copper, pure....................	212	222		
Gold....................	64	169		
Gold....................	212	170		
Iron, pure....................	64	39.0		
Iron, pure....................	212	36.6		
Iron, wrought....................	64	34.9		
Iron, wrought....................	212	34.6		
Iron, cast....................	32	29.0		
Iron, cast....................	212	28.0		
Steel, mild....................	32	36	36.2	−0.016
Steel, mild....................	212	33		
Steel, 1% C....................	64	26.2		
Steel, 1% C....................	212	25.9		
Lead....................	32	20.0		
Lead....................	212	19.8		
Magnesium....................	32–212	92.0		
Mercury....................	32	4.8		
Mercury....................	122	4.8		
Platinum....................	64	40.2		
Platinum....................	212	41.9		
Silver....................	32	244		
Silver....................	212	240		
Tin....................	32	36		
Tin....................	212	34		
Tungsten....................	68	92.5		
Zinc....................	32	65		
Zinc....................	212	62		

* Mainly from *International Critical Tables*.

TABLE 2-1. THERMAL CONDUCTIVITIES OF SOLIDS* (*Continued*)

Substance	Temperature, °F	k	k_0	α
a. Metals (*Continued*):				
Alloys:				
Bearing metal, white..............	68	13.7		
Brass, 90–10....................	32	59	57.5	0.049
Brass, 90–10....................	212	68		
Brass, 60–40....................	32	61	59	0.052
Brass, 60–40....................	212	69		
Bronze........................	68	33.6		
Bronze........................	210	40		
Constantan, 60 Cu; 40 Ni........	64	13.1		
Constantan, 60 Cu; 40 Ni........	212	15.5		
Manganin, 84 Cu; 4 Ni; 12 Mn....	64	12.8		
	212	15.2		
Nickel alloy: 70 Ni; 28 Cu; 2 Fe...	68	20.2		
62 Ni; 12 Cr; 26 Fe............	68	7.8		
Nickel silver....................	32	16.9		
Nickel silver....................	212	21.5		
Platinoid......................	64	14.5		
b. Structural and heat-resistant materials:				
Asphalt.......................	68–132	0.43–0.44		
Bricks:				
Building brick, common.........	68	0.40		
Building brick, face.............		0.76		
Carborundum brick.............	1110	10.7		
Carborundum brick.............	2550	6.4		
Chrome brick..................	392	1.34		
Chrome brick..................	1022	1.43		
Chrome brick..................	1652	1.15		
Diatomaceous earth, molded and				
fired.......................	400	0.14		
	1600	0.18		
Fire-clay brick (burnt 2426°F)...	932	0.60		
	1472	0.62		
	2012	0.63		
Fire-clay brick (burnt 2642°F)...	932	0.74		
	1472	0.79		
	2012	0.81		

* Mainly from *International Critical Tables.*

TABLE 2-1. THERMAL CONDUCTIVITIES OF SOLIDS* (*Continued*)

Substance	Temperature, °F	k	k_0	α
b. Structural and heat-resistant materials, Bricks (*Continued*):				
Fire-clay brick (Missouri)......	392	0.58		
	1112	0.85		
	2552	1.02		
Magnesite.................	400	2.2		
	1200	1.6		
	2200	1.1		
Cement, Portland..............		0.17		
Cement mortar.................	75	0.67		
Concrete, cinder................	75	0.44		
Concrete, stone 1–2–4 mix........	69	0.79		
Glass, window.................	68	Av. 0.45		
Glass, borosilicate..............	86–167	0.63		
Plaster, gypsum................	70	0.28		
Plaster, metal lath..............	70	0.27		
Plaster, wood lath..............	70	0.16		
Stones:				
Granite.......................		1.0–2.3		
Limestone.....................	210–570	0.73–0.77		
Marble.......................		1.20–1.70		
Sandstone....................	104	1.06		
Wood (across the grain):				
Balsa, 8.8 lb/cu ft..............	86	0.032		
Cypress.......................	86	0.056		
Fir...........................	75	0.063		
Maple or oak..................	86	0.096		
Yellow pine...................	75	0.085		
White pine....................	86	0.065		
c. Insulating materials:				
Asbestos:				
Asbestos cement boards........	68	0.43		
Asbestos sheets...............	124	0.096		
Asbestos felt, 40 laminations per inch....................	100	0.033		
	300	0.040		
	500	0.048		
Asbestos felt, 20 laminations per inch....................	100	0.045		
	300	0.055		
	500	0.065		

* Mainly from *International Critical Tables.*

TABLE 2-1. THERMAL CONDUCTIVITIES OF SOLIDS* (*Continued*)

Substance	Temperature, °F	k	k₀	α
c. Insulating materials, Asbestos (*Continued*):				
Asbestos, corrugated, 4 plies per inch......................	100	0.05		
	200	0.058		
	300	0.069		
Asbestos cement.................	......	1.2		
Asbestos, loosely packed........	−50	0.086		
	32	0.089		
	210	0.093		
Balsam wool, 2.2 lb/cu ft.........	90	0.023		
Cardboard, corrugated.............	......	0.037		
Celotex.........................	90	0.028		
Corkboard, 10 lb/cu ft............	86	0.025		
Cork, regranulated...............	90	0.026		
Cork, ground....................	90	0.025		
Diatomaceous earth (Sil-o-cel).....	32	0.035		
Felt, hair.......................	86	0.021		
Felt, wool......................	86	0.03		
Fiber insulating board............	70	0.028		
Glass wool, 1.5 lb/cu ft...........	75	0.022		
Insulex, dry....................	90	0.037–		
		0.083		
Kapok..........................	86	0.020		
Magnesia, 85%.................	100	0.039		
	200	0.041		
	300	0.043		
	400	0.046		
Rock wool, 10 lb/cu ft............	90	0.023		
Rock wool, loosely packed........	300	0.039		
	500	0.050		
Sawdust........................	75	0.034		
Silica aerogel...................	90	0.014	0.012	
Wood shavings..................	75	0.034		

* Mainly from *International Critical Tables*.

of metals and alloys can be predicted. There are several equations[4] relating the thermal conductivity to the electrical conductivity. Most of these equations have very limited ranges of application, and the accuracy is not good, especially for the alloys. Figures 2-2 and 2-3 show that for some metals and alloys the value of k at a temperature t may be determined by the straight-line relationship

$$k = k_0 + \alpha t \qquad \text{Btu/(hr)(sq ft)(°F/ft)} \qquad (2\text{-}2)$$

where k_0 = thermal conductivity at 0°F and α = constant that depends upon the material. If k increases with temperature, then α is positive; whereas if k decreases with temperature, then α is negative. Values of α are included in Table 2-1 for those materials for which it has been determined. In the case of those materials for which no value of α is given, either it is not possible to include it, owing to insufficient experimental data, or else k is not a straight-line function of t. It must be remembered that the values of k in Table 2-1 for metals are accurate only for the compositions given. In general, slight amounts of impurities in the pure metals make great changes in the value of k. For the alloys, however, slight changes in the percentages of the alloying constituents have little effect on the value of k.

There are many factors which influence the thermal conductivities of the structural and heat-resisting materials. Factors such as density and moisture contents, as well as composition, crystal structure, and temperature, will affect the thermal conductivities of these materials. In the case of refractory materials, the temperature at which they are fired will have a sizable effect. For wood, in addition to the factors just mentioned, the direction of the grain will have a very large effect on k. Because of the influences of the factors mentioned, the values given in Table 2-1 for structural and heat-resistant materials should be used for purposes of estimates rather than exact values. There is much material in the literature showing these effects. However, for exact values for a particular material, it is usually necessary to measure the value of k.

2-5. Thermal Conductivity of Insulating Materials. Any material or device which offers a high resistance to the transfer of heat by conduction, radiation, or convection may serve as a form of insulation. A good insulating material, in addition to high resistance to the transfer of heat, must have various characteristics, depending upon the application. In some types of service, such as furnace insulation, the material must be able to withstand high temperatures without deterioration. In aircraft insulation some structural strength, an ability to withstand vibration, and light weight are essential. Insulating materials should be odorless and should not absorb odors. They should withstand rot or disintegration and should not provide a dwelling place or be a food for rodents and insects. This is particularly true for insulation of buildings and food-storage plants. All insulating materials should either be unaffected by moisture or should be thoroughly protected against the entrance of moisture.

Of the various forms of insulating materials, by far the majority utilize the characteristic of low thermal conductivity as the means of restricting the transfer of heat, although radiation and convection are also significant. Restriction of heat transfer by radiation is applied in some forms of commercial insulation.

In general, insulating materials may be considered to consist of small air spaces surrounded by solid walls. The low thermal conductivity of such materials may be attributed to the low thermal conductivity of the air enclosed within the interstices or cells of the material and the relatively small area of solid material through which the heat may be conducted.

In general, for low temperature differences on the two sides of the insulation, those materials of lower apparent density will show the lower thermal conductivities. For larger temperature differences, convection currents may arise within the air spaces and reduce the insulating value of such materials. Indeed, for such a situation, a somewhat denser structure may give better results. For all such materials the higher the operating temperature the higher will be the conductance, since radiation through the gas will play a more important role, and there will also be an increase in conductance due to internal convection.

2-6. Thermal Conductivity of Fluids. An inspection of Tables 2-2 and 2-3, showing the thermal conductivities of a number of gases and liquids, respectively, indicates that at ordinary temperatures the value of k for nonmetallic pure liquids lies in the range from 0.05 Btu/(hr)(sq ft)(°F/ft) to approximately 0.40, while for gases the range of k is from 0.005 to 0.015. Thus

TABLE 2-2. THERMAL CONDUCTIVITIES OF GASES AND VAPORS*
$k = $ Btu/(hr)(sq ft)(°F/ft)

Substance	Temperature, °F	k	k_{32}	C
Air† (see Table A-2)	−312 to 414		0.0141	225
	0	0.0132		
	200	0.0182		
	1000	0.0362		
Ammonia, NH_3	32	0.0116	0.0116	
	212	0.0171		
Carbon dioxide, CO_2	−109	0.0053	0.0709	
	212	0.0119		
	540	0.0164		
	1031	0.0343		
Freon-11‡	86	0.0048		
	194	0.0056		
Freon-12‡	86	0.0056		
	194	0.0070		
Freon-21‡	86	0.0057		
	194	0.0063		
Freon-22‡	86	0.0068		
	194	0.0080		
Freon-113‡ (at ½ atm)	86	0.0045		
	194	0.0059		
Freon-114‡	86	0.0065		
	194	0.0081		
Hydrogen, H_2	−422 to 212		0.0917	169
Methylene chloride, CH_2Cl_2 (Carrene No. 1)	32	0.0035	0.0035	
	124	0.0044		
	212	0.0058		
	415	0.0085		
Methyl chloride, CH_3Cl	32	0.0049	0.0049	
	124	0.0066		
	212	0.0086		
	363	0.0119		
Nitrogen, N_2	−312 to 212		0.0131	205
Oxygen, O_2	−312 to 212		0.0134	259
Sulfur dioxide, SO_2	32	0.0045	0.0045	
Water vapor, H_2O	115	0.0112		
	212	0.0139		

* Mainly from *International Critical Tables*, McGraw-Hill Book Company, Inc., New York, 1926–1933.
† From *Gas Tables* by J. H. Keenan and J. Kaye, John Wiley & Sons, Inc., New York, 1948.
‡ Data supplied by Kinetic Chemicals Div., E. I. du Pont de Nemours & Co., Wilmington, Del.

TABLE 2-3. THERMAL CONDUCTIVITIES OF LIQUIDS*
$k = Btu/(hr)(sq\ ft)(°F/ft)$

Substance	Temperature, °F	k	k_0	α
Ammonia, NH_3 (26%)	32 176	0.242 0.318		
Ammonia (average)	14 to 68	0.29		
Brine (25% NaCl)	−4 176	0.274 0.368		
Carbon dioxide, CO_2	68	0.121		
Freon-11†	32 167	0.0680 0.0503		
Freon-12†	32 167	0.0559 0.0392		
Freon-21†	32 167	0.0770 0.0590		
Freon-22†	32 104	0.0704 0.0559		
Freon-113†	32 167	0.0576 0.0440		
Freon-114†	32 167	0.0515 0.0344		
Kerosene	68 176	0.0872 0.0795		
Mercury	32 212	4.83 4.83		
Methyl chloride, CH_3Cl	68	0.093	0.1125	−0.000297
Methylene chloride, CH_2Cl_2 (Carrene No. 1)	68	0.099	0.112	−0.000192
Oil: Castor Light heat transfer Rabbeth spindle	102 86 212 86 212	0.104 0.0765 0.0748 0.0825 0.0805		
Sulfur dioxide, SO_2	68	0.115	0.1285	−0.000211
Water (see Table A-1)	32 212	0.327 0.394		

* Mainly from *International Critical Tables*.
† Data supplied by Kinetic Chemicals Div., E. I. du Pont de Nemours & Co., Wilmington, Del.

the largest value of k for gases is less than one-third of the smallest value of k for liquids.

Several attempts have been made to derive the theoretical and empirical equations to relate the thermal conductivity of liquids to the other physical properties.[5] In general, these equations are applicable over a limited range, with a fairly large error possible in many instances. The variation of the value of k with temperature for several liquids may be expressed by Eq. (2-2), with the constant α given in Table 2-3 for those cases where it is known to apply. In most cases the thermal conductivity of liquids decreases with increase in temperature. For water, however, which is the best conductor among the nonmetallic liquids, the value of α in Eq. (2-2) is positive up to a temperature of 240°F, which indicates an increase in k with increase in temperature. For temperatures above 280°F the thermal conductivity of water decreases as t increases; hence k decreases from 0.396 at 280°F to 0.356 at 500°F.

The liquid metals, as well as some aqueous solutions, have thermal conductivities higher than the value for water. Liquid mercury, for example, has a k of 4.83 Btu/(hr)(sq ft)(°F/ft). Some other examples are molten aluminum with a k of 51 and molten lead with a k of 8.7.

The thermal conductivity of a gas will, in general, increase with increase in temperature and is virtually independent of the pressure, provided it is not too far from atmospheric conditions. The thermal conductivity of a gas mixture is not equal to the sum of the products of the individual values of k multiplied by the percentages of each constituent, although, for very rough approximations, such a rule may be used. The heavier gases have lower thermal conductivities than the lighter gases.

From purely theoretical considerations Maxwell[6] deduced, from the kinetic theory of gases, an expression for the relation of the thermal conductivity of a gas to its viscosity. This equation, which is supported to a remarkable degree by experimental data, is

$$k = a\mu c_v \quad \text{Btu/(hr)(sq ft)(°F/ft)} \tag{2-3}$$

where μ = absolute viscosity, $\text{lb}_m/(\text{ft})(\text{hr})$.

c_v = specific heat at constant volume, $\text{Btu/(lb}_m)(°F)$.

a = constant having the following theoretical values:

2.45 for monatomic gases (He, A).

1.90 for diatomic gases (air, N_2, O_2, CO, H_2).

1.70 for triatomic gases (CO_2, H_2O).

1.30 for more complex gases.

Experiments show that the value of a for diatomic gases may vary from 1.803 to 1.95. For gases having more than three atoms, values of a from 1.385 to 1.41 have been measured.

Eucken[7] combined Eq. (2-3) with the Sutherland equation for variation of viscosity with temperature, which is

$$\mu_t = \mu_{32} \frac{492 + C}{T + C} \left(\frac{T}{492} \right)^{3/2} \qquad lb_m/(ft)(hr) \qquad (2-4)$$

where μ_t = absolute viscosity at the temperature t.

μ_{32} = absolute viscosity at 32°F.

T = absolute temperature, °F.

C = constant depending on the gas.

Eucken's work shows that the resulting equation, which is

$$k = k_{32} \frac{492 + C}{T + C} \left(\frac{T}{492} \right)^{3/2} \qquad Btu/(hr)(sq\ ft)(°F/ft) \qquad (2-5)$$

is true for several gases at temperatures within the ranges noted in Table 2-2. Where experimentally determined values are lacking, Eq. (2-5) is usually applied beyond the temperature range indicated, but with doubtful accuracy. In Eq. (2-5), k is the coefficient of thermal conductivity at the temperature t, T is the absolute temperature in degrees Fahrenheit ($T = t + 460$), and C is a constant depending on the gas. Values of C are included in Table 2-2 for those gases to which Eq. (2-5) is known to be applicable. Values of μ may be found in Table A-5 and Fig. A-7.

2-7. Probable Accuracy of Thermal-conductivity Coefficients.
It should be remembered that the values of k given in Tables 2-1, 2-2, and 2-3 represent the best available information at the present time. The accuracy of the values given for homogeneous solid materials is probably quite high, but the values for non-homogeneous solid materials are subject to considerable uncertainty because of variation in composition. This is especially true of porous substances whose moisture content may vary materially.

Because the effect of radiation and convection heat transfer was not completely eliminated in determining the thermal conductivity of some of the gases reported in Table 2-2, some of the values may possibly be in error as much as 10 to 20 per cent.

The thermal conductivities shown for liquids in Table 2-3 are subject to errors similar to those for gases, although the accuracy may be somewhat better.

In calculating the heat transfer by conduction through a material, the value of k should be taken at the mean temperature. In some cases this may be calculated by the formula given in Art. 2-4. Where this is not possible, and the value of k is given for two or more temperatures, it is probably sufficiently accurate to assume a straight-line variation of the thermal conductivity with the temperature, if the temperature variation is not too large.

Table 2-4 shows the order of magnitude of k for various classes of materials. These values may be used for very rough estimates of the heat transfer by conduction.

TABLE 2-4. ORDER OF MAGNITUDE OF k FOR VARIOUS CLASSES
OF MATERIALS

Material	k	
Gases	0.005–	0.02
Insulating materials	0.014–	0.10
Wood	0.04 –	0.10
Liquids (nonmetallic)	0.05 –	0.40
Brick, concrete, stone, plaster	0.2 –	2.0
Refractory materials	0.50 –	10.0
Metals and alloys	10	–240

Problems

1. Calculate the thermal conductivity of a 4- by 4-ft test panel, 0.80 in. thick, if during a 3-hr test period there is conducted 900 Btu through the panel with a temperature differential of 15°F between surfaces.

2. Electric current is supplied to a 9- by 9-in. guarded hot plate (similar to the one shown in Fig. 2-1) at the rate of 10 watts, maintaining a temperature of 130°F on the hot side of two test specimens of solid material, each 1 in. thick, with the cold sides at 70°F. What is the value of k for the material at the mean temperature of 100°F, expressed in Btu/(hr)(sq ft)(°F/ft)?

3. Methyl chloride at a pressure of 62 psia boils at 60°F. What is its thermal conductivity k in Btu/(hr)(sq ft)(°F/ft) at the pressure of 62 psia (a) if its temperature is 30°F, and (b) if its temperature is 62°F?

4. Calculate the thermal conductivity k in Btu/(hr)(sq ft)(°F/ft) for sulfur dioxide gas at 160°F if the specific heat at constant volume is 0.123 Btu/(lb)(°F) and the viscosity is as shown in Fig. A-7.

5. What is the thermal conductivity of oxygen at 0°F? Compare the answer with that obtained from Eq. (2-3) if c_v for oxygen at 0°F is 0.155 and the viscosity is as shown in Fig. A-7.

6. Calculate the thermal conductivity of air at 1000°F by application of Eucken's equation and the data of Table 2-2.

7. Calculate the thermal conductivity of nitrogen at a mean temperature of 180°F.

REFERENCES

1. Fourier, J. B.: "Théorie analytique de la chaleur," Gauthier-Villars, Paris, 1822; English translation by Freeman, Cambridge, 1878.
2. Jakob, M.: "Heat Transfer," vol. 1, pp. 146–166, John Wiley & Sons, Inc., New York, 1949.
3. American Society for Testing Materials: Standard method of test for thermal conductivity of materials by means of the guarded hot plate, C 177–45, Philadelphia, 1945.
4. Jakob, M.: "Heat Transfer," vol. 1, pp. 112–117, John Wiley & Sons, Inc., New York, 1949.
5. Smith, J. F. D.: The Thermal Conductivity of Liquids, *Trans. ASME.* **58**, 719 (1936).
6. Maxwell, J. C.: "Collected Works," vol. II, p. 1, Cambridge, London, 1890.
7. Eucken, A., *Physik. Z.*, **14**, 324 (1913).

CHAPTER 3

EQUATIONS FOR THE CALCULATION OF CONDUCTION HEAT TRANSFER

STEADY STATE

3-1. General. For heat flow when the temperatures at all points do not vary with time, the temperature and heat flow may be calculated when the conductivity, temperature conditions at the surfaces, and the shape are known. Exact mathematical solutions have been worked out for a few regular shapes, but approximations must be resorted to in many practical cases. In any event it is necessary to integrate Eq. (2-1) between the proper limits before it can be used for the solution of heat-transfer problems.

Equation (2-1) for the general case may be integrated as follows:

$$q = -kA \frac{dt}{dL}$$

$$q \int_{L_1}^{L_2} \frac{dL}{A} = - \int_{t_1}^{t_2} k \, dt \tag{3-1}$$

The variation of the thermal conductivity with temperature may be expressed by the functional relationship

$$k = f(t) \tag{3-2}$$

and Eq. (3-1) may be written

$$q \int_{L_1}^{L_2} \frac{dL}{A} = - \int_{t_1}^{t_2} f(t) \, dt \tag{3-3}$$

If the right-hand member of Eq. (3-3) is multiplied and divided by the temperature difference $t_2 - t_1$, Eq. (3-3) becomes

$$q \int_{L_1}^{L_2} \frac{dL}{A} = \frac{- \int_{t_1}^{t_2} f(t) \, dt}{t_2 - t_1} (t_2 - t_1) \tag{3-4}$$

But the term

$$\frac{\int_{t_1}^{t_2} f(t)\,dt}{t_2 - t_1}$$

is the mean value of $f(t)$ between t_1 and t_2, which in this case is equal to k_m, the mean value of k over this temperature range; therefore

$$q \int_{L_1}^{L_2} \frac{dL}{A} = -k_m(t_2 - t_1)$$

or

$$q = \frac{k_m(t_1 - t_2)}{\int_{L_1}^{L_2} (dL/A)} \tag{3-5}$$

Hence, without error, the term $- \int_{t_1}^{t_2} k\,dt$ has been replaced by $k_m(t_1 - t_2)$ irrespective of the relation between A and L.

3-2. Conduction through a Single Plane Wall (Unidirectional Flow). In the case of a single plane wall, as shown in Fig. 3-1, with parallel surfaces, x ft apart, maintained at constant uniform temperatures t_1 and t_2, the area A is constant and independent of the distance dL. Here, L_1 is zero and L_2 is x; hence the integration of Eq. (3-5) yields the formula

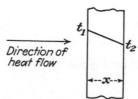

Fig. 3-1. Single plane wall.

$$q = \frac{A k_m(t_1 - t_2)}{x} \quad \text{Btu/hr} \tag{3-6}$$

where A = area, sq ft, taken perpendicular to the direction of flow.

k_m = mean thermal conductivity, Btu/(hr)(sq ft)(°F/ft) for the temperature range.

$t_1 - t_2$ = temperature difference, °F.

Equation (3-6) applies to any homogeneous body having a constant cross section in which heat flows only in one direction (i.e., perpendicular to the surfaces). It may be used to calculate the heat conduction in a plane wall if its length and width, perpendicular to the direction of heat flow, are large compared to the dimension in the direction of heat flow, thus making the conduction through the edges small enough to neglect. Equation (3-6) may also be used in the case of axial heat conduction through a cylindrical or prismatic bar with parallel ends maintained at temperatures t_1 and t_2 while the rest of the bar is insulated to prevent flow of heat through the sides.

For any case in which the cross-sectional area varies with L it is necessary to substitute for A, in Eq. (3-5), its value in terms of L.

3-3. Thermal Resistance and Thermal Conductance. In Art. 2-2 it was implied that heat conduction and electrical conduction may be considered as analogous processes. This analogy may be considered to apply to the determination of the over-all thermal resistance or conductance of circuits, consisting of several thermal conductors, by the method used to determine the over-all electrical resistance or conductance of circuits consisting of several electrical conductors. To this end it is to be noted that Eq. (3-5) expresses the rate of heat transfer in a form similar to Ohm's law for the flow of electrical current; i.e.,

$$q = \frac{t_1 - t_2}{\frac{1}{k_m} \int_{L_1}^{L_2} \frac{dL}{A}} \tag{3-7}$$

is analogous to

$$I = \frac{E}{R_e} \tag{3-8}$$

where the heat-transfer rate q corresponds to the current-flow rate I, the temperature difference $t_1 - t_2$ corresponds to the voltage difference E, and the term $(1/k_m) \int_{L_1}^{L_2} dL/A$ corresponds to the electrical resistance R_e. From this viewpoint the thermal resistance would be

$$R_t = \frac{1}{k_m} \int_{L_1}^{L_2} \frac{dL}{A} \qquad (°F)(hr)/Btu \tag{3-9}$$

It is apparent that the *thermal resistance* of any thermal conductor is the ratio of the temperature drop across that conductor to the heat-transfer rate through it. Similarly, the *thermal conductance* may be defined by the equation

$$C_t = \frac{1}{R_t} \qquad Btu/(°F)(hr) \tag{3-10}$$

and is the ratio of the heat-transfer rate to the temperature drop.

In keeping with this analogy, therefore, it is to be expected that the over-all thermal resistance of conductors in series will be given by the equation

$$R_{t_{(series)}} = R_{t_1} + R_{t_2} + \cdots + R_{t_n} \qquad (°F)(hr)/Btu \quad (3\text{-}11)$$

while the over-all thermal resistance of conductors in parallel is

$$R_{t_{(parallel)}} = \frac{1}{\dfrac{1}{R_{t_1}} + \dfrac{1}{R_{t_2}} + \cdots + \dfrac{1}{R_{t_n}}} \qquad (°F)(hr)/Btu \quad (3\text{-}12)$$

The heat-transfer rate as given by Eq. (3-7) may be written

$$q = \frac{t_1 - t_2}{R_t} \qquad (3\text{-}13)$$

where for a single plane wall, from Eq. (3-6), it is seen that

$$R_t = \frac{x}{k_m A} \qquad (°F)(hr)/Btu \qquad (3\text{-}14)$$

Equations (3-13) and (3-14) may be used to determine heat-transfer rates, thermal resistances, and temperature drops in some composite walls.

3-4. Conduction through Composite Walls (Unidirectional Conduction). Figure 3-2 shows sections of three types of

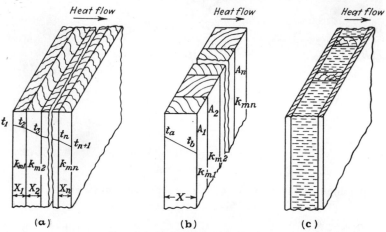

FIG. 3-2. Composite plane walls.

composite walls. A wall made up of slabs in series is shown in a, a wall with parallel slabs is shown in b, and one with both parallel and series slabs is shown in c.

In the series composite wall of Fig. 3-2a the thermal conductivities are k_{m1}, k_{m2}, . . . , k_{mn}, the thicknesses are x_1, x_2, . . . , x_n, and the temperatures on the faces of the individual slabs, in the direction of heat conduction, are t_1, t_2, t_3, . . . , t_{n+1}. Since steady state is assumed, the heat-transfer rate q through an area A of any slab (designated by the subscript i) is exactly equal to the heat-transfer rate through the same area A of any other slab. Thus by Eq. (3-13)

$$q = \frac{t_i - t_{i+1}}{R_{ti}} \tag{3-15}$$

and by Eq. (3-14)

$$R_{ti} = \frac{x_i}{k_{mi}A} \tag{3-16}$$

Equation (3-15) may be rearranged to read

$$qR_{ti} = t_i - t_{i+1} \tag{3-17}$$

and both sides may be summed up to give the total temperature drop from $i = 1$ to $i = n$. Thus

$$q \sum_{i=1}^{i=n} R_{ti} = \sum_{i=1}^{i=n} (t_i - t_{i+1}) \tag{3-18}$$

or

$$q(R_{t1} + R_{t2} + \cdots + R_{tn}) = (t_1 - t_2) + (t_2 - t_3) + \cdots \\ + (t_n - t_{n+1}) \tag{3-19}$$

which, when Eq. (3-11) is substituted, gives

$$qR_t = t_1 - t_{n+1} \tag{3-20}$$

or

$$q = \frac{t_1 - t_{n+1}}{R_t} \quad \text{Btu/hr} \tag{3-21}$$

Equations (3-11) and (3-14) may be used to show that

$$R_t = \frac{x_1}{k_{m1}A} + \frac{x_2}{k_{m2}A} + \cdots + \frac{x_n}{k_{mn}A} \quad \text{(°F)(hr)/Btu} \tag{3-22}$$

By dividing Eq. (3-17) by Eq. (3-20) the temperature drop across the slab i is found to be

$$t_i - t_{i+1} = \frac{R_{ti}}{R_t} (t_1 - t_{n+1}) \qquad (3\text{-}23)$$

This shows that the ratio of temperature drop across any slab to the over-all temperature drop is proportional to the ratio of the thermal resistance of the slab to the over-all thermal resistance.

In the case of the composite wall composed of slabs in parallel (Fig. 3-2b), the left-hand surface is maintained at a temperature t_a and the right-hand surface is maintained at a temperature t_b. The thermal conductivities of the slabs are k_{m1}, k_{m2}, ... , k_{mn}, and the areas of the slabs, perpendicular to the direction of heat flow, are A_1, A_2, ... , A_n.

For steady state, the heat-transfer rate through the slab i is

$$q_i = \frac{t_a - t_b}{R_{ti}} \qquad (3\text{-}24)$$

The total rate of heat transfer through the wall is found by taking the sum of the q_i's from $i = 1$ to $i = n$. Thus,

$$\sum_{i=1}^{i=n} q_i = (t_a - t_b) \sum_{i=1}^{i=n} \frac{1}{R_{ti}} \qquad (3\text{-}25)$$

or

$$q = \frac{t_a - t_b}{R_t} \quad \text{Btu/hr} \qquad (3\text{-}26)$$

From Eq. (3-12)

$$R_t = \frac{1}{\dfrac{1}{R_{t1}} + \dfrac{1}{R_{t2}} + \cdots + \dfrac{1}{R_{tn}}} \qquad (3\text{-}27)$$

By using Eq. (3-14) the total resistance is found to be

$$R_t = \frac{x}{k_{m1}A_1 + k_{m2}A_2 + \cdots + k_{mn}A_n} \quad (\text{°F})(\text{hr})/\text{Btu} \quad (3\text{-}28)$$

The heat-transfer equations for a wall such as the one shown in Fig. 3-2c may be developed with methods similar to those used for the series and parallel walls by dividing the wall into a set of parallel slabs, each composed of slabs in series.

The use of the equations developed for composite walls is restricted to walls in which the heat flow is in one direction only (infinite slabs). They may be used for real walls, however, if the dimensions perpendicular to the direction of heat flow are large compared to the dimensions in the direction of heat flow.

3-5. Conduction through Thick-walled Cylinders (Two-dimensional or Radial Conduction). The thermal resistance of the hollow cylinder shown in Fig. 3-3 may be determined by the use

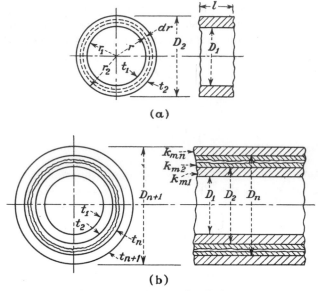

(a)

(b)

FIG. 3-3. Thick-walled cylinders.

of Eq. (3-9). In this case the temperature of the inner surface is t_1 and its radius is r_1. The temperature of the outer surface is $t_2 < t_1$, and its radius is r_2. Equation (3-9) may be applied to an annulus of length l, between the inner and outer surface having an area $A = 2\pi r l$ and a thickness $dL = dr$. When this expression is integrated from $L_1 = r_1$ to $L_2 = r_2$ the thermal resistance is

$$R_t = \frac{1}{k_m} \int_{r_1}^{r_2} \frac{dr}{2\pi l r}$$

or $$R_t = \frac{1}{2\pi l k_m} \ln \frac{r_2}{r_1} \qquad (°F)(hr)/Btu \qquad (3\text{-}29)$$

If this value of R_t is substituted into Eq. (3-13), and if the diameter ratio D_2/D_1 is substituted for r_2/r_1, the heat-transfer rate is

$$q = \frac{2\pi l k_m(t_1 - t_2)}{\ln (D_2/D_1)} \qquad \text{Btu/hr} \qquad (3\text{-}30)$$

The hollow cylinder of Fig. 3-3a could be treated as a plane wall having a thickness $x = (D_1 - D_2)/2$ and a heat-transfer area $A = (\pi/2)(D_2 + D_1)l$, provided the diameter ratio D_1/D_2 is greater than 0.6. If these values for x and A are substituted into Eq. (3-6), the result is

$$q = \frac{\pi(D_2 + D_1)l k_m(t_1 - t_2)}{D_2 - D_1} \qquad \text{Btu/hr} \qquad (3\text{-}31)$$

The heat-transfer rate computed by Eq. (3-31) is in error by an amount less than 2.20 per cent for $0.6 < D_1/D_2 < 1$.

The over-all thermal resistance for the composite cylinder shown in Fig. 3-3b, made up of concentric cylinders having thermal conductivities $k_{m1}, k_{m2}, \ldots, k_{mn}$, is found by using Eq. (3-11), since this represents a series system. The heat-transfer rate for any cylinder i may be written

$$q = \frac{t_i - t_{i+1}}{R_{t_i}} \qquad (3\text{-}32)$$

where, from Eq. (3-29),

$$R_{ti} = \frac{1}{2\pi l k_{mi}} \ln \frac{r_{i+1}}{r_i} \qquad (3\text{-}33)$$

If the diameter ratio is substituted for the radius ratio, the result substituted in Eq. (3-32), and this result summed from $i = 1$ to $i = n$, the heat-transfer rate is

$$q = \frac{2\pi l(t_1 - t_{n+1})}{\dfrac{1}{k_{m1}} \ln \dfrac{D_2}{D_1} + \dfrac{1}{k_{m2}} \ln \dfrac{D_3}{D_2} + \cdots + \dfrac{1}{k_{mn}} \ln \dfrac{D_{n+1}}{D_n}} \qquad \text{Btu/hr}$$

$$(3\text{-}34)$$

As in the case of the composite plane wall forming a series system, the temperature drop across any cylinder in this system may be found by the use of Eq. (3-23).

3-6. Conduction through Various Practical Shapes (Two- and Three-dimensional Conduction). Frequently in practice steady-state conduction problems involving shapes other than plane walls or insulated prismatic bars are encountered. Such multi-dimensional heat-conduction problems may be handled by several methods. These include mathematical, numerical, and graphical methods along with empirical formulas. Several of these methods are described in Chaps. 13 and 14.

The results of some approximate solutions for the case of hollow rectangular parallelepipeds are included in this chapter because they have rather wide application to practical configurations such as rectangular-shaped furnaces with thick walls.

Langmuir's[1] empirical equations for several important special cases are given below. These equations are applicable only to hollow rectangular parallelepipeds, where corresponding inner and outer surfaces are parallel and in all cases the same distance L apart. It is also assumed that the temperatures t_1 and t_2 are measured at the inside and outside surfaces of the walls and that they are uniform on all sides.

CASE I.   The length of each inside edge is between one-fifth and twice the thickness L. If A is the total area of the inside faces in square feet, L the thickness in feet, and Σe the sum of the lengths of all inside edges in feet, then the rate of heat flow by conduction is

$$q = \left(\frac{A}{L} + 0.54\Sigma e + 1.2L\right) k_m(t_1 - t_2) \qquad \text{Btu/hr} \quad (3\text{-}35)$$

CASE II.   The length of one inside dimension is less than one-fifth the thickness L. If Σe is the sum of the remaining eight inside edges, each greater than one-fifth L, and all other symbols are the same as for Case I, then

$$q = \left(\frac{A}{L} + 0.465\Sigma e + 0.35L\right) k_m(t_1 - t_2) \qquad \text{Btu/hr} \quad (3\text{-}36)$$

CASE III.   The lengths of two inside dimensions are each less than one-fifth of the thickness L. If e is the length of the longest inside edge in feet, and A_1 is the total area of the outer faces in square feet, then

$$q = \frac{2.78e}{\log_{10}{(A_1/A)}} k_m(t_1 - t_2) \qquad \text{Btu/hr} \qquad (3\text{-}37)$$

CASE IV. For small, hollow, rectangular parallelepipeds where all three inside dimensions are less than one-fifth the thickness L,

$$q = \frac{0.79 \sqrt{AA_1}}{L} k_m(t_1 - t_2) \qquad \text{Btu/hr} \qquad (3\text{-}38)$$

where all symbols are as already defined.

3-7. Summary of Steady-state Heat Conduction. The equations in this chapter apply only to steady-state heat-conduction problems and can be used only if the temperatures at the outside surfaces are known. In many cases it is not possible to determine these temperatures, but it is possible to determine the temperatures of the fluid in contact with the solid surfaces. For such cases the conductance of the fluid film must be determined. This film conductance is usually designated by the letter h and is defined as the quantity of heat that will be conducted through 1 sq ft of the film in 1 hr for each degree Fahrenheit of temperature difference between the temperatures of the main body of the fluid and of the solid surface with which it is in contact. The value of h is dependent on many different factors which are discussed in later chapters.

In Eq. (3-5) it is seen that the conduction is dependent upon k_m, the mean value of the coefficient of thermal conductivity between the temperatures t_1 and t_2. In some cases this value may be calculated by the formulas given in Chap. 2. In most practical cases, however, where the temperature difference $t_1 - t_2$ is small, it is sufficiently accurate to use the arithmetic mean of k_1 at temperature t_1 and k_2 at temperature t_2.

In using Eqs. (3-22) and (3-34) it is necessary to know the temperatures between the layers before the values of k_{m1}, k_{m2}, etc., can be determined accurately. However, most practical problems of conduction through composite walls fall into one of several types for which certain simplifications may be made that make it possible to calculate the approximate heat conduction without very great error. These classifications and simplifications are as follows:

1. Single wall of high conductivity (metal) in series with a single wall of low conductivity (insulating material). In this case little error is made if k for the metal wall is evaluated at the temperature of its exposed surface and k for the insulation is con-

sidered as the arithmetic mean k for the over-all temperature range.

2. Single metal wall of high conductivity followed by two layers of insulating material. For problems of this type the value of k for the metal wall is evaluated at the temperature of its exposed surface and the arithmetic mean value of k for the insulating layers may be determined by approximating the temperature drop through each layer. By this method a first approximation of the heat conduction is obtained that is usually sufficiently accurate for most practical purposes. If greater accuracy is sought, successive approximations will yield the desired result. However, it should be remembered that more than one approximation may be unwarranted in view of the limited accuracy of the tabulated values of k.

Several practical examples are given here to illustrate the methods suggested above.

a. Simple Plane Wall (Both Surface Temperatures Known). The interior-wall temperature of an annealing oven constructed of 10-in. firebrick (Missouri) is 2552°F, and the exterior-wall temperature is 392°F. What is the heat loss per hour per square foot?

Since this is a simple plane wall, Eq. (3-6) should be used.

$$q = \frac{A k_m (t_1 - t_2)}{x}$$

Table 2-1 gives the following values of k for various temperatures:

t, °F	k, Btu/(hr)(sq ft)(°F/ft)
392	0.58
1112	0.85
2552	1.02

The arithmetic mean value of k for the temperatures 2552 and 392°F is

$$k_m = \frac{0.58 + 1.02}{2} = 0.80 \text{ Btu/(hr)(sq ft)(°F/ft)}$$

Therefore

$$q = \frac{1 \times 0.80(2552 - 392)}{10/12}$$
$$= 2074 \text{ Btu/(hr)(sq ft)}$$

If the true mean value of k were determined by integration, the value of q would be 2230 Btu/(hr)(sq ft). Thus it is seen that for this case the heat loss calculated with the arithmetic mean k is approximately 7 per cent too low.

This particular example was chosen because k for this material varies approximately as the one-third power of the temperature. If k varied as the one-half power of the temperature, the calculated heat loss would still be too low but the error would be smaller (of the order of $3\frac{1}{2}$ per cent); whereas if k varied as the first power of the temperature (the most common case for homogeneous solids), the arithmetic mean value of k would be the true mean value and no error would be involved.

b. *Conduction through Composite Plane Wall (Both Surface Temperatures Known).* A heat exchanger made of $\frac{1}{4}$-in. steel plate is covered with 3 in. of 85 per cent magnesia insulation. If the temperature of the inside steel surface is 300°F and the temperature of the insulation at the outer surface is 100°F, what is the heat loss in Btu/(hr)(sq ft)? For a composite plane wall with two layers Eqs. (3-21) and (3-22) may be used with $n = 2$. Thus

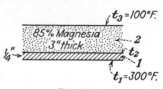

Fig. 3-4

$$q = \frac{A(t_1 - t_3)}{(x_1/k_{m1}) + (x_2/k_{m2})} \quad \text{Btu/hr}$$

This is a composite wall of the type discussed under 1 of this article. The value of k_{m1} for the steel is obtained from Fig. 2-2 or Table 2-1 at the temperature of its exposed surface. Here it is seen that k for steel at 300°F is 31.4 Btu/(hr)(sq ft)(°F/ft). Table 2-1 shows that for 85 per cent magnesia, k at the mean temperature of 200°F is 0.041.

The heat transfer by conduction would be

$$q = \frac{1(300 - 100)}{0.25/(12 \times 31.4) + 3/(12 \times 0.041)} = 32.8 \text{ Btu/(hr)(sq ft)}$$

It can be seen that the temperature drop across the steel wall is very small; i.e., from Eq. (3-13)

$$t_1 - t_2 = qR_{t1} \qquad °F$$
$$300 - t_2 = \frac{32.8 \times 0.25}{1 \times 31.4 \times 12}$$
$$300 - t_2 = 0.02$$

or
$$t_2 = 299.98°F$$

Therefore the error involved, if it is assumed that the temperature of the unexposed surface of the insulation is the same as the temperature of the exposed metal surface, is extremely small.

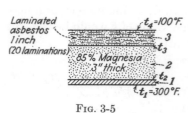

Laminated asbestos 1 inch (20 laminations)

$t_4 = 100°F$.

85% Magnesia 3" thick

$t_1 = 300°F$.

FIG. 3-5

c. *Composite Plane Wall with Three Layers (Both Surface Temperatures Known).* If the heat exchanger of example b is covered with 1 in. of laminated asbestos felt (20 laminations to the inch) over the other insulation, what will be the heat loss when the surface temperatures are the same as before? For this problem Eqs. (3-21) and (3-22) may be used.

$$q = \frac{A(t_1 - t_4)}{(x_1/k_{m1}) + (x_2/k_{m2}) + (x_3/k_{m3})} \qquad \text{Btu/hr}$$

This is the type of wall discussed under 2 of this article. As in the previous problem, k_{m1} for the steel is 31.4 Btu/(hr)(sq ft) (°F/ft). A first approximation of the temperature drop through each layer of insulation must be made before proper values of k_{m2} and k_{m3} can be found. The temperature drop through a layer is inversely proportional to the conductance of the material, i.e., directly proportional to the resistance x/k_m. It may be assumed for the first approximation that k_{m2} and k_{m3} are the arithmetic mean values of the coefficients at 100 and 300°F; then k_{m2} is found from Table 2-1 to be 0.041, and k_{m3} is found to be 0.050. A first approximation of the temperature drop across each layer may be found by the method of Art. 3-4. Since the resistance of layer 1 is extremely small it may be neglected, and the temperature at the interface between layers 1 and 2 may be assumed to be 300°F. The total temperature drop of 200°F is therefore assumed to occur across the layers 2 and 3. The resistance R_{t2} of layer 2 is $x_2/k_{m2} = \frac{3}{12} \div 0.041 = 6.1$, and the resistance R_{t3} of layer 3 is $x_3/k_{m3} = \frac{1}{12} \div 0.050 = 1.67$. Thus R_t, the total resistance, is $R_{t2} + R_{t3} = 6.1 + 1.67 = 7.77$. The temperature drop across layer 2 will be

$$\Delta t_2 = (R_{t2}/R_t)\,\Delta t = (6.1/7.77) \times 200 = 157°F$$

and the temperature drop across layer 3 is

$$\Delta t_3 = (R_{t3}/R_t)\,\Delta t = (1.67/7.77) \times 200 = 43°F$$

This first approximation then shows that the temperature t_3 between the second and third layers will be approximately

$$300 - 157 = 143°F$$

By interpolation in Table 2-1, the value of k at this temperature for the 85 per cent magnesia is found to be practically 0.04, whereas k for the laminated asbestos at this temperature is 0.047. From this table the value of k for 85 per cent magnesia at 300°F is found to be 0.043 and k for the asbestos at 100°F is 0.045; hence,

$$k_{m2} = \frac{0.04 + 0.043}{2} = 0.0415 \text{ Btu/(hr)(sq ft)(°F/ft)}$$

and

$$k_{m3} = \frac{0.047 + 0.045}{2} = 0.046 \text{ Btu/(hr)(sq ft)(°F/ft)}$$

By substitution of these values in the equation,

$$q = \frac{1(300 - 100)}{0.25/(12 \times 31.4) + 3/(12 \times 0.0415) + 1/(12 \times 0.046)}$$
$$= 24.6 \text{ Btu/(hr)(sq ft)}$$

If Eq. (3-23) is applied to layer 3, then the temperature drop across the laminated asbestos is found to be approximately 44.6°F instead of 43°F as calculated in the first approximation. This small error in temperature drop would have practically no effect on the values of k_{m2} and k_{m3}; hence it would not affect the heat conduction. Thus it is seen that for this case a first approximation is sufficient.

It should be noticed in examples a and b that the steel wall, which has a high coefficient of thermal conductivity, has no appreciable effect on the amount of heat conduction. This would still be true even if the steel wall were ten times as thick as it actually was in this case.

d. Conduction through Insulated Pipes (Inside- and Outside-surface Temperatures Known). A 3-in. standard steel pipe is covered with 3 in. of 85 per cent magnesia. If the temperature on the inside surface is 300°F and the temperature on the outside surface is 100°F, what is the rate of heat transfer in Btu/(hr)(ft of pipe)?

For this example Eq. (3-34) may be used with $n = 2$. This becomes

$$\frac{q}{l} = \frac{2\pi(t_1 - t_3)}{\dfrac{1}{k_{m1}} \ln \dfrac{D_2}{D_1} + \dfrac{1}{k_{m2}} \ln \dfrac{D_3}{D_2}} \qquad \text{Btu/(hr)(ft)}$$

The same assumption can be made here that was made in exam-

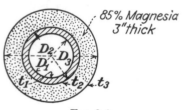

FIG. 3-6

ple *b*; viz., the temperature drop through the metal wall is zero. Then k_{m1} for the steel pipe wall at 300°F is 31.4. This value is found in Table 2-1. For the 85 per cent magnesia the two surface temperatures are 100 and 300°F. Table 2-1 shows the value of k for the mean temperature of 200°F to be 0.041. Hence the rate of heat transfer is

$$\frac{q}{l} = \frac{2\pi(300 - 100)}{\dfrac{1}{31.4} \ln \dfrac{3.5}{3.07} + \dfrac{1}{0.041} \ln \dfrac{9.5}{3.5}}$$
$$= 51.8 \text{ Btu/(hr)(ft)}$$

The heat transfer in Btu/(hr)(sq ft of outer surface) may be found by dividing this result by the outer area of a 1-ft length of the pipe, as in the following calculation:

$$\frac{q}{A_3} = \frac{51.8}{\pi \times (9.5/12)} = 20.8 \text{ Btu/(hr)(sq ft)}$$

A comparison of this result with that obtained in the case of the composite plane wall (example *b*) shows that for the same temperatures, materials, and thicknesses, the heat lost through the curved wall is, in this case, 36.6 per cent less than that lost through the plane wall.

Problems

1. What is the rate of heat transfer through a piece of celotex 3 ft by 8 ft by 1 in. in thickness, if the temperature of one surface is 80°F and of the other surface is 60°F?

2. A metal rod 1 sq in. in cross section and 6 in. long is heated at one end and cooled at the other. If the rate of heat input is 3 Btu/hr and the difference in temperature of the two ends is 2°F, what is the coefficient of thermal conductivity in Btu/(hr)(sq ft)(°F/ft)?

3. A horizontal plate of steel 2 in. thick is covered by a blanket of insulation of the same thickness. The temperature of the lower side of the steel is 500°F and of the upper side of the insulation 100°F. k for steel equals 29; k for the insulation equals 1.0. Determine the temperature of the upper side of the steel plate.

4. A solid wall is made up of face brick 4 in. thick, cement mortar ½ in. thick, and 1–2–4 mix stone concrete 8 in. thick. If the temperature of the exposed surface of the concrete is 70°F and the temperature of the exposed surface of the face brick is 31°F, find (a) the heat transfer in Btu/(hr) (sq ft), and (b) the temperature between the mortar and the concrete.

5. A composite wall is made up of common brick 8 in. thick, faced with sandstone 6 in. thick. The room-side surface of the brick is plastered with gypsum plaster ¾ in. thick. The temperature of the exposed surface of the plaster is 90°F. Determine the temperature of the exposed surface of the sandstone and of both surfaces of the brick when the rate of heat flow from the room side to the outside is 20 Btu/(hr)(sq ft).

6. A furnace wall consists of 9 in. of Missouri firebrick, 6 in. of common brick, 2 in. of 85 per cent magnesia, and ⅛ in. steel plate on the outside. If the inside-surface temperature is 2400°F and the outside-surface temperature is 200°F, estimate the temperatures between layers and calculate the heat loss in Btu/(hr)(sq ft). Assume straight-line extrapolation of values of k for 85 per cent magnesia at temperatures of 100 and 400°F.

7. A plane wall is composed of an 8-in. layer of refractory brick ($k = 0.75$) and a 2-in. layer of insulating material with k for the insulating material varying linearly as $k = 0.02 + 0.0001t$, where t is the temperature in degrees Fahrenheit. The inside-surface temperature of the brick is 2000°F, and the outside-surface temperature of the insulating material is 100°F. Calculate the temperature at the boundary of the brick and insulation.

8. An insulated wall is to be constructed of common brick 8 in. thick and metal lath and plaster 1 in. thick, with an intermediate layer of loosely packed rock wool. The outer surfaces of the brick and plaster are to be at temperatures of 1000 and 120°F, respectively. Calculate the thickness of insulation required in order that the heat loss per square foot shall not exceed 100 Btu/hr.

9. A composite wall is made of fire-clay brick (Missouri) 8 in. thick and 85 per cent magnesia insulation. The temperature of the exposed surface of the fire-clay brick is 800°F and that of the exposed surface of the insulation is 100°F. (a) What thickness of insulation is needed to provide a temperature at the interface not exceeding 700°F? (b) What does the temperature become if the thickness of insulation is doubled? (c) What is the rate of heat transfer in each case?

10. A floor is composed of common brick 4 in. thick laid on a 4-in. slab of 1–2–4 mix stone concrete which is laid directly on the ground. The thermal conductivity of the ground is 0.90 Btu/(hr)(sq ft)(°F/ft). Calculate the temperature of the ground at a depth of 12 in. below the lower side of the slab when the temperatures of the upper and lower surfaces of the brick are 70 and 66°F, respectively. Assume the heat flow to be normal to the surface of the floor.

11. A refrigerator, insulated with corkboard 2 in. thick, maintains an average inside-surface temperature of 35°F when the outside-surface temperature is 65°F and provides a usable space of 20 by 20 by 36 in. Determine the percentage increase in usable space for the same outside dimensions if the corkboard is replaced on all six sides by silica aerogel of sufficient thickness to maintain the same rate of heat flow. Neglect the thickness and resistance of the enclosing metal and consider the surface as a single plane wall equivalent to the outside surface area.

12. A furnace wall is made of two layers of refractory material each 9 in. thick. The surface temperatures on the furnace side of layer a and on the outer side of layer b are 2000 and 200°F, respectively. For layer a, $k = 2.44 - 0.0006t$, and for layer b, $k = 0.06 + 0.00009t$, where t is the mean temperature of the layer. Determine (a) the heat flow through the wall in Btu/(hr)(sq ft), (b) the temperature at the junction of layers a and b, and (c) the heat flow through the wall if layers a and b are interchanged.

13. The temperature of the outside surface of an 8-in. double extra-strong bare steel pipe (i.d. = 6.87 in. and o.d. = 8.625 in.) is 595°F, and the temperature of the inside surface is 600°F. (a) What is the heat loss per foot of pipe? (b) What will be the saving in heat if the pipe is insulated with 2 in. of 85 per cent magnesia and the temperature of the outer surface of the insulation is reduced to 200°F?

14. A 6-in. nominal-diameter mild steel pipe (i.d. = 6.065 in. and o.d. = 6.625 in.) has an inside-surface temperature of 500°F. The pipe is covered with a 2-in. layer of 85 per cent magnesia insulation which has a surface temperature of 200°F. Determine (a) the heat loss in Btu/hr per 100 ft of pipe, and (b) the percentage error which occurs if the arithmetic mean area is used instead of the logarithmic mean area.

15. A corkboard box has outside dimensions of 20 by 22 by 24 in., and the walls are 6 in. thick. The temperature of the inside surface is −20°F, and the temperature of the outside surface is 60°F. Find the rate at which heat must be removed from the inside of the box in order to maintain these temperatures.

16. A cast-iron box having outside dimensions of 36 by 36 by 26 in. has a wall thickness of 12 in. If the inside-surface temperature is maintained at 300°F and the outside-surface temperature is 100°F, what is the heat loss in Btu/hr?

17. A hollow sphere has an inside-surface temperature of 500°F and an outside-surface temperature of 50°F. If $k = 10$, calculate the heat loss by conduction for an inside diameter of 2 in. and an outside diameter of 6 in. The surface of a sphere is $4\pi r^2$. Apply Eq. (3-5).

18. What would be the heat loss by conduction if the equation for a plane wall, of area equal to the mean of the inside and outside surfaces, is assumed to apply to the sphere in the above problem?

19. Two layers of yellow pine, each 1 in. thick, are held tightly together by many small mild steel nails 2 in. long which extend through the wood, flush with the outer surfaces. The cross-sectional area of the nails is 1 per cent of the total cross-sectional area through which heat travels. Calculate (a) the rate of heat transfer in Btu/(hr)(sq ft) for temperatures of 90 and 70°F on the two exposed surfaces, and (b) the total conductance C in Btu/(hr)(sq ft)(°F).

20. Repeat Prob. 19a but with a 1-in. layer of gypsum plasterboard in contact with each of the two surfaces which were initially exposed. The temperatures of 90 and 70°F are now to apply to the exposed surfaces of the plasterboard. Note that the resistance R of the yellow pine and nails equals $1/C$.

21. A hollow sphere is heated by means of an internal heating coil having a resistance of 100 ohms. If the mean thermal conductivity of the sphere material is 30 Btu/(hr)(sq ft)(°F/ft), calculate the current necessary to maintain a temperature difference between the inside and outside surfaces of 8°F. The inside and outside diameters of the sphere are 8 and 9 in., respectively.

22. Two cylindrical metal bars of diameter d and lengths l_1 and l_2, having mean thermal conductivities k_1 and k_2, respectively, are perfectly insulated on the circumference. If the right-hand face of bar 1 is in contact with the left-hand face of bar 2, and if the left-hand face of bar 1 is maintained at a temperature t_1 while the right-hand face of bar 2 is maintained at a temperature t_3, establish expressions for (a) the total thermal resistance, and (b) the temperature t_2 at the interface.

23. Two concentric cylinders are made of materials having thermal conductivities k_1 and k_2 for the inner and outer cylinders, respectively. The inside cylinder has radii r_1 and r_2 at its inside and outside surfaces, while the outside cylinder has radii r_2 and r_3 at its inside and outside surfaces. The temperature at the radius r_1 is maintained at t_1, while the temperature at the radius r_3 is maintained at t_3. Establish expressions for (a) the thermal resistance for unit length, and (b) the temperature t_2 at the interface.

REFERENCE

1. Langmuir, I., E. Q. Adams, and G. S. Meikle, *Trans. Am. Electrochem. Soc.*, **24**, 53 (1913).

CHAPTER 4

RADIATION

4-1. General. In the first chapter it was stated that heat may be transmitted from one body to another without altering the temperature of the intervening medium. That this is true may be proved by a simple, everyday experience. If one stands before a fire or a hot radiator, he experiences a sensation of warmth that is not due to the temperature of the air; for if a screen is interposed, the sensation immediately disappears. This would not be the case if the air had a high temperature. This phenomenon, which is called thermal radiation, is but one of the many forms of radiant energy that is continuously being emitted and absorbed in various degrees by all bodies. All radiant energy may be regarded as a form of wave motion, known as an electromagnetic phenomenon. This type of wave motion occurs in the "ether"* and should not be confused with sound waves, spring vibrations, and other elastic-mechanical waves which may occur in solids, liquids, and gases only, whereas radiant-energy waves may be transmitted even through a vacuum.

4-2. Spectral-energy Distribution. A given body under a fixed set of conditions will emit radiation of various wavelengths.† The amount or intensity of the radiation in the various wavelengths is different. The type of radiation emitted from a certain

---

* At one time all radiation was believed to be electromagnetic waves of various wavelengths carried by the "ether." At present the theory relating to radiation that has found great favor among scientists is the quantum theory; however, the true nature of radiation and the nature of the carrier of radiant energy have not been established completely.

† Radiation waves travel with a velocity V that depends upon the medium through which they are propagated. The frequency of radiation ν is dependent only on the source; it does not vary with the medium. The ratio $V/\nu = \lambda$ is called the wavelength of radiation and depends upon the medium through which it is propagated as well as the source.

body is characterized by the band of wavelengths having the greatest intensity.

The curves of Fig. 4-1 show the distribution of intensity of radiation with wavelength for a "black body"* at various temperatures. These curves are from the measurements of Lummer

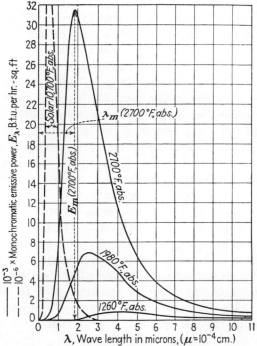

FIG. 4-1. Energy distribution of a black body.

and Pringsheim.[1] The wavelength scale is in microns;† the temperatures are in degrees Fahrenheit absolute; and the emissive power is in Btu per hour per micron for 1 sq ft of surface. The following conclusions may be drawn from these curves:

1. An increase in temperature T causes a decrease in λ_m, the wavelength at which maximum energy emission occurs. The rate of energy emission at λ_m is called E_m.

* A black body is one that emits the maximum possible radiation at a given temperature. The adjective "black" has nothing to do with the color of the body. This is discussed in greater detail on pp. 49–50.

† One micron is 10^{-4} cm, or 3.937×10^{-5} in. The symbol μ indicates one micron.

2. An increase in temperature causes a rapid increase in energy emission at any given wavelength.

3. The total rate of energy emission at any temperature and for any range of wavelengths is given by the area under the curve for that temperature taken over the wavelength range being considered.

The differentiation among various types of radiation, such as light radiation, thermal radiation, etc., is rather indefinite, since radiation of all wavelengths will heat bodies. It is evident from the curves of Fig. 4-1 that practically all the radiation of a black body at 2700°F abs lies in the range of wavelengths from 0.6 to 20μ, which is mostly in the invisible infrared-radiation region. Visible radiation lies in the range of wavelengths from 0.4 to 0.8μ. Solar radiation, assuming the temperature of the sun to be 10,240°F, lies within the range of wavelengths from 0.1 to 3μ; the greater portion, however, lies in the visible region.

4-3. Factors in Thermal Radiation. In 1792 Prevost proposed the "theory of exchanges," which states that there is a continuous interchange of energy among bodies as a result of the reciprocal process of radiation and absorption. Thus, if two bodies at different temperatures are within an enclosure, the hotter body receives, from the colder body, less energy than it radiates; consequently its temperature decreases; whereas the colder body receives more energy than it radiates, and its temperature increases. This interchange of energy continues even after thermal equilibrium is reached, except that both bodies then receive as much energy as they radiate. According to this concept, which agrees well with observations, any body would cease to emit *thermal* radiation only when its temperature has been reduced to absolute zero.

It has been shown experimentally that the higher the temperature of a body becomes the faster it radiates heat energy. Obviously the amount of heat radiated per unit of time will also be proportional to the amount of surface exposed. It has also been found that two bodies alike as to material, size, weight, and temperature, but with different surface finishes, will have quite different rates of radiation. The amount of radiation that a body will absorb depends upon its temperature, amount of surface, surface finish, and the angle at which the rays strike the surface. Good absorbers of heat (when cold) are generally also good radiators of heat (when hot).

The relationships among the foregoing factors involved in thermal radiation are given by a number of radiation formulas; some have been deduced from theoretical considerations and proved to be true by experiment, whereas others are purely empirical. To understand these formulas, their use and limitations, it is necessary to understand some of the fundamental laws and definitions of radiation theory.

4-4. Emission of Radiation. It has been stated that the emission of radiation is common to all bodies and that the amount of radiant energy emitted per unit time is dependent upon the nature of the body, its temperature, and the kind and extent of its surface. The exact process by which nature converts heat energy into radiant energy or radiant energy into heat energy is not known. It is thought that radiation requires for its emission and absorption something that, in behavior characteristics, simulates within the atom some sort of oscillator. Although the exact mechanism is not known, the laws of radiation have been formulated with some precision.

Every particle within and on the surface of a body emits radiant energy. It would seem, therefore, that the amount of radiant energy emitted from a body would depend upon its volume as well as the other properties previously mentioned. Observations prove that this is not entirely true. The explanation is that radiation emitted by the particles at some distance from the surface of a body is absorbed by other parts of the body before it reaches the surface. Only the radiation originating at the surface or at a limited distance beneath the surface can escape from the body. The distance beneath the surface from which radiation can escape or to which radiation can penetrate is limited by the transparency of the body. The term transparency is generally used to indicate the ease with which light can pass through a body; however, in this instance it is used to indicate the ease with which all thermal radiation can pass through a substance. Some substances are transparent to radiation of certain wavelengths and opaque to radiation of other wavelengths that may be emitted from the same source. Most solids are opaque to nearly all thermal radiation, and therefore the emission (or absorption) of radiation takes place within a very thin surface layer. A notable exception to this is glass which, although a solid, is transparent to short-wavelength thermal radiation (light) but is opaque to longer-wavelength radiation emitted by bodies at any temperature lower than that required to produce light. Liquids and gases as well as some other solids are transparent to a greater or lesser degree.

4-5. Absorption, Reflection, and Transmission of Radiation. When radiant energy falls upon a body, part or all of it may be absorbed, part or all of it may be reflected, and part or all of it may be transmitted (pass through the body undiminished).

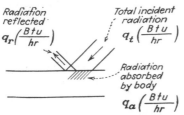

FIG. 4-2. Distribution of radiation on an opaque body.

An opaque body will absorb and reflect all the radiation falling upon it. Figure 4-2 illustrates the way in which radiation is distributed when it strikes the surface of an opaque body. From the illustration it can be seen that

$$q_a + q_r = q_t \tag{4-1}$$

or if

$$\frac{q_a}{q_t} = a \quad \text{and} \quad \frac{q_r}{q_t} = r$$

then

$$a + r = 1 \tag{4-2}$$

In Eq. (4-2) $a = absorptivity$, or the fraction of the incident radiation absorbed, and $r = reflectivity$, or the fraction of the incident radiation reflected at the surface. It can be seen from Eq. (4-2) that a can be increased or decreased by changing r; i.e., the amount of radiation absorbed by an opaque body will be changed if the reflectivity is changed. The reflectivity is dependent upon the character of the surface. Hence, the amount

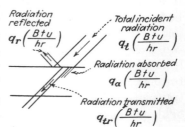

FIG. 4-3. Distribution of radiation on a transparent body.

of radiation absorbed by an opaque body may be increased or decreased by appropriate surface treatment.

When radiation falls upon a transparent body, as indicated in

Fig. 4-3, the radiation that is absorbed depends upon the transmissivity as well as the reflectivity; i.e.,

$$q_a + q_r + q_{tr} = q_t \tag{4-3}$$

or if $\qquad \dfrac{q_a}{q_t} = a, \qquad \dfrac{q_r}{q_t} = r \qquad$ and $\qquad \dfrac{q_{tr}}{q_t} = t_r$

then $\qquad\qquad\qquad a + r + t_r = 1 \tag{4-4}$

Here t_r, the transmissivity, is the fraction of the incident radiation that is transmitted completely through the body, and it depends upon the thickness of the body as well as upon the reflectivity r and absorptivity a.

4-6. Definitions and Fundamental Laws. *a. Total Emissive Power.* The total emissive power of a body, designated by the symbol E, is the total radiant energy emitted per unit time per unit area of radiating surface. In this text the units of E will be Btu/(hr)(sq ft). From Fig. 4-1 it can be seen that E, for any particular temperature, is the area under the curve, for that temperature, from $\lambda = 0$ to $\lambda = \infty$, or

$$E = \sum_{\lambda=0}^{\lambda=\infty} E_\lambda \, d\lambda \tag{4-5}$$

where E_λ is defined as the *monochromatic emissive power*, i.e., the radiant energy emitted by the body at a particular temperature and wavelength. For a body that gives off radiation at all wavelengths, i.e., where E_λ is a continuous function of λ, Eq. (4-5) becomes

$$E = \int_0^\infty E_\lambda \, d\lambda \tag{4-6}$$

The solution of Eq. (4-5) or (4-6) is very difficult for actual bodies encountered in practice. For this reason the properties of a black body must be investigated before attempting a practical solution of Eq. (4-6).

b. The Black Body and Its Properties. There are, in nature, no bodies for which the reflectivity r is zero or for which the absorptivity a is unity; i.e., no body absorbs all the radiant energy falling upon it. Some bodies, such as lampblack and platinum black, reflect only a very small fraction of the incident radiation, and consequently they are called "black," although the term has little to do with the color, in the optical sense. The perfect black body is one whose surface absorbs all the radiant energy incident upon it.   For such a body $a = 1$ and $r = 0$.

Although there is no body in nature for which $a = 1$, one has been assumed for theoretical considerations and approximated very closely for experiments by which the theory has been checked.   In all experimental determinations of black-body radiation a small hole in the side of a hollow enclosure is used. In Fig. 4-4 a ray of radiant energy entering the enclosure will be

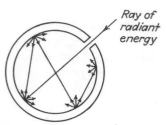

partially absorbed as it strikes the inside surface.   The remainder will be *diffusely* reflected, i.e., reflected in all directions.   The reflected energy will strike other parts of the interior and will be partially absorbed.   Only a very small portion of the diffusely reflected energy will find its way out through the hole; the rest will be completely absorbed by successive

FIG. 4-4. Hollow enclosure for measuring radiation.

reflections.   Theoretically perfect absorption will take place only when the area of the hole is infinitely small compared with the internal area of the enclosure.   Practically, an approximation sufficiently accurate for experimental purposes is obtained by using a hole ½ in. in diameter in the end of a hollow cylindrical tube 8 in. long and 2 in. in diameter.

Upon heating an enclosure, such as has been described, the inside walls radiate energy and some of this radiation passes out through the hole.   (The curves of Fig. 4-1 show the radiation from just such a body.)   The energy streaming from this hole is called "black-body" radiation.

Numerous investigators,[2] during the first decade of the twentieth century, attempted to derive a theoretical equation showing the relationship between E_λ and λ to fit the experimental curves of Fig. 4-1.   All deductions that were based on statistical mechanics and classical thermodynamics* led to formulas that only partially agreed with observed values.   In 1900, however,

* Among the formulas proposed were Wien's formula, $E_\lambda = c_1\lambda^{-5}e^{-C_2/\lambda T}$ [see *Ann. Physik*, **58,** 662 (1896)], and Rayleigh's formula,

$$E_\lambda = c_1\lambda^{-4}Te^{-C_2/\lambda T}$$

[see *Phil. Mag.*, **49,** 539 (1900)].

Planck[3] introduced the quantum theory, which enabled him to express the relationship between E_λ and λ in such a manner that it exactly fits the experimental values.

Planck's equation has the form

$$E_\lambda = \frac{1.16 \times 10^8 \lambda^{-5}}{e^{25740/\lambda T} - 1} \qquad \text{Btu/(sq ft)(hr)}(\mu) \qquad (4\text{-}7)$$

when expressed in consistent units,

where E_λ = monochromatic emissive power of a black body, Btu/(sq ft)(hr)(μ).

  λ = wavelength, μ.

  T = temperature of the radiating black body, °F abs.

  e = Napierian base of logarithms which is numerically equal to 2.718.

Planck's formula has been checked experimentally and theoretically and is accepted as an exact relationship among E, λ, and T for black-body radiation.

When the right-hand term of Eq. (4-7) is multiplied by $d\lambda$ and integrated* between the limits $\lambda = 0$ to $\lambda = \infty$, it yields the Stefan-Boltzmann† law, which is

$$E_B = 0.173 \times 10^{-8} \times T^4 \qquad \text{Btu/(sq ft)(hr)} \qquad (4\text{-}8)$$

This law states that the total energy radiated by a perfect black body is proportional to the fourth power of the absolute temperature. The areas under the curves of Fig. 4-1 represent the values obtained by use of Eq. (4-8).

  c. Relation between Absorptivity and Emissive Power. A little investigation will show that there is a definite relationship between the total emissive power E of any body at a given temperature and its absorptivity a at the same temperature. Suppose, for example, that a small body of a given shape and surface area S is placed in a hollow, evacuated iron sphere kept at a constant, uniform, absolute temperature T. After a time the body within the sphere will reach the same temperature as

---

* See appendix for mathematical development.

† The Stefan-Boltzmann law was arrived at about 20 years before Planck proposed his formula. In 1879 Stefan, on the basis of experimental data observed by Tyndall, was led to suggest that the total emissive power of *any* body is proportional to the fourth power of its absolute temperature. Subsequently Boltzmann deduced the same law from theoretical considerations. See *Wiedemann's Ann.*, **22**, 31, 291 (1884).

that of the interior and thereafter will radiate as much energy as it absorbs. Let the radiation falling on the body per unit time and per unit area be I, and let E_1 and a_1 be the respective total emissive power and the absorptivity of the body. Since the temperature of the body remains constant,

$$E_1 S = I a_1 S$$
or
$$E_1 = I a_1$$

Now suppose that the first body is replaced, in exactly the same position, by a second body of the same surface area and shape but of entirely different material. After the second body attains the temperature T of the enclosure, the equilibrium equation may be written

$$E_2 = I a_2$$

Similarly if the second body is replaced by a perfect black body of the same surface area and shape, then

$$E_B = I a_B$$

It is to be noted that the radiation I falling per unit time and per unit area on each body, in its turn, is exactly the same, since the enclosure, its temperature, and the size and shape of the bodies remain the same. Now from the equations for these three bodies,

$$\frac{E_1}{a_1} = \frac{E_2}{a_2} = \frac{E_B}{a_B} \tag{4-9}$$

But for a black body $a_B = 1$; hence

$$\frac{E_1}{a_1} = \frac{E_2}{a_2} = E_B \tag{4-10}$$

This may be written

$$E_1 = a_1 E_B \tag{4-11}$$
$$E_2 = a_2 E_B$$

In words: *At a given temperature the total emissive power for any body is equal to its absorptivity multiplied by the total emissive power of a perfect black body at that temperature; i.e.,*

$$E = a E_B \tag{4-12}$$

This is known as Kirchhoff's law, derived by Kirchhoff in 1859, based upon experiments performed by Ritchie in 1833.

The ratio of the total emissive power of any body to the total emissive power of a black body, at the same temperature, is called the *emissivity* ε and is numerically equal to the absorptivity:

$$\frac{E}{E_B} = a = \epsilon \tag{4-13}$$

or $\qquad\qquad E = \epsilon E_B \qquad \text{Btu}/(\text{sq ft})(\text{hr}) \tag{4-14}$

When Eq. (4-14) is combined with Eq. (4-8), it yields the expression

$$E = \epsilon \times 0.173 \times 10^{-8}T^4 \qquad \text{Btu}/(\text{sq ft})(\text{hr}) \tag{4-15}$$

These expressions permit the calculation of the total emissive power of any body if the temperature T and the emissivity ε are known. The normal emissivities* of various materials at several temperatures are given in Table 4-1. These values are the experimental results obtained by many investigators. Where two temperatures and two emissivities are shown in the same horizontal line, the first temperature corresponds to the first value of emissivity, the second to the second, and linear interpolation is permitted. In general, the emission from the surface of any body falls short of the corresponding black-body emission at all wavelengths, and the distribution curve therefore lies below that for a black body at the same temperature. A surface whose emissivity is the same at all wavelengths and temperatures is called a "gray" surface and has emission curves similar to those of a black body, but reduced in intensity in constant ratio. The radiation from a gray surface, therefore, may be calculated *exactly* by Eq. (4-15), since ε does not vary with T. The ratio of the radiation of an iron body to the radiation of a black body is different at different wavelengths. For many such surfaces that are not gray, the mean value of ε at the temperature being

* Normal emissivity is the emissivity for radiation in a direction normal to the surface from which energy is being emitted. The emissivity of a surface will vary with direction. In general, the average emissivity for all directions is only slightly different from the normal emissivity. This difference is largest for very highly polished metallic surfaces where the average emissivity may be as much as 20 per cent larger than the normal emissivity and is smallest for rough surfaces where the average emissivity is approximately the same as the normal emissivity. For most practical purposes the normal-emissivity values given in Table 4-1 may be used in place of the average emissivity.

TABLE 4-1. THE NORMAL TOTAL EMISSIVITY OF VARIOUS SURFACES
(*From Hottel*[6])

Surfaces	°F	ϵ
A. Metals and Their Oxides		
Aluminum:		
Highly polished plate, 98.3% pure..........	440, 1070	0.039, 0.057
Polished plate............................	73	0.040
Rough plate..............................	78	0.055
Oxidized at 1110°F.......................	390, 1110	0.11, 0.19
Al-surfaced roofing.......................	110	0.216
Al-treated surfaces, heated at 1110°F:		
Copper................................	390, 1110	0.18, 0.19
Steel.................................	390, 1110	0.52, 0.57
Brass:		
Highly polished:		
73.2% Cu, 26.7% Zn, by weight..........	476, 674	0.028, 0.031
62.4% Cu, 36.8% Zn, 0.4% Pb, 0.3% Al,		
by weight...........................	494, 710	0.0388, 0.037
82.9% Cu, 17.0% Zn, by weight..........	530	0.030
Hard-rolled, polished, but direction of polishing		
visible...........................	70	0.038
But somewhat attacked..................	73	0.043
But traces of stearin from polish left on....	75	0.053
Polished...............................	100, 600	0.096, 0.096
Rolled plate:		
Natural surface........................	72	0.06
Rubbed with coarse emery...............	72	0.20
Dull plate..............................	120, 660	0.22
Oxidized by heating at 1110°F..............	390, 1110	0.61, 0.59
Chromium:		
See Nickel Alloys for Ni-Cr steels		
Copper:		
Carefully polished electrolytic Cu..........	176	0.018
Commercial, emeried, polished, but pits re-		
maining.............................	66	0.030
Scraped shiny, but not mirrorlike.........	72	0.072
Polished................................	242	0.023
Plate heated at 1110°F....................	390, 1110	0.57, 0.57
Cuprous oxide...........................	1470, 2010	0.66, 0.54
Plate, heated for a long time, covered with thick		
oxide layer............................	77	0.78
Molten copper...........................	1970, 2330	0.16, 0.13
Gold:		
Pure, highly polished.....................	440, 1160	0.018, 0.035

Table 4-1. The Normal Total Emissivity of Various Surfaces.
(*Continued*)

Surfaces	°F	ϵ
Iron and steel:		
Metallic surfaces (or very thin oxide layer):		
Electrolytic iron, highly polished...........	350, 440	0.052, 0.074
Polished iron...........................	800, 1880	0.144, 0.377
Iron freshly emeried.....................	68	0.242
Cast iron, polished......................	˙392	0.21
Wrought iron, highly polished.............	100, 480	0.28
Cast iron, newly turned..................	72	0.435
Polished steel casting....................	1420, 1900	0.52, 0.56
Ground sheet steel......................	1720, 2010	0.55, 0.61
Smooth sheet iron.......................	1650, 1900	0.55, 0.60
Cast iron, turned on lathe................	1620, 1810	0.60, 0.70
Oxidized surfaces:		
Iron plate, pickled, then rusted red........	68	0.612
Then completely rusted................	67	0.685
Rolled sheet steel.......................	70	0.657
Oxidized iron...........................	212	0.736
Cast iron, oxidized at 1100°F..............	390, 1110	0.64, 0.78
Steel oxidized at 1100°F..................	390, 1110	0.79, 0.79
Smooth, oxidized electrolytic iron..........	260, 980	0.78, 0.82
Iron oxide..............................	930, 2190	0.85, 0.89
Rough ingot iron........................	1700, 2040	0.87, 0.95
Sheet steel, strong rough oxide layer........	75	0.80
Dense shiny oxide layer................	75	0.82
Cast plate:		
Smooth..............................	73	0.80
Rough...............................	73	0.82
Cast iron, rough, strongly oxidized........	100, 480	0.95
Wrought iron, dull-oxidized...............	70, 680	0.94
Steel plate, rough.......................	100, 700	0.94, 0.97
High-temperature alloy steels; see Nickel alloys		
Molten metals:		
Molten cast iron........................	2370, 2550	0.29, 0.29
Molten mild steel.......................	2910, 3270	0.28, 0.28
Lead:		
˙ Pure (99.96%) unoxidized..................	260, 440	0.057, 0.075
Gray oxidized...........................	75	0.281
Oxidized at 390°F.......................	390	0.63
Mercury, pure clean......................	32, 212	0.09, 0.12
Molybdenum filament......................	1340, 4700	0.096, 0.292
Ni-Cu alloy, oxidized at 1110°F...............	390, 1110	0.41, 0.46

TABLE 4-1. THE NORMAL TOTAL EMISSIVITY OF VARIOUS SURFACES.
(*Continued*)

Surfaces	°F	ε
Nickel:		
Electroplated on polished iron, then polished..	74	0.045
Technically pure (98.9% Ni by weight, + Mn),		
polished..............................	440, 710	0.07, 0.087
Electroplated on pickled iron, not polished....	68	0.11
Wire...................................	368, 1844	0.096, 0.186
Plate, oxidized by heating at 1110°F........	390, 1110	0.37, 0.48
Nickel oxide.............................	1200, 2290	0.59, 0.86
Nickel alloys:		
Cr-Ni alloy..............................	125, 1894	0.64, 0.76
(18–32% Ni, 55–68% Cu, 20% Zn by weight),		
gray oxidized..........................	70	0.262
Alloy steel (8% Ni, 18% Cr); light silvery,		
rough; brown after heating...............	420, 914	0.44, 0.36
Same, after 24 hr heating at 980°F...........	420, 980	0.62, 0.73
Alloy (20% Ni, 25% Cr), brown, splotched,		
oxidized from service.....................	420, 980	0.90, 0.97
Alloy (60% Ni, 12% Cr), smooth, black, firm		
adhesive oxide coat from service...........	520, 1045	0.89, 0.82
Platinum:		
Pure, polished plate......................	440, 1160	0.054, 0.104
Strip..................................	1700, 2960	0.12, 0.17
Filament...............................	80, 2240	0.036, 0.192
Wire...................................	440, 2510	0.073, 0.182
Silver:		
Polished, pure...........................	440, 1160	0.0198, 0.0324
Polished................................	100, 700	0.0221, 0.0312
Steel, see Iron		
Tantalum filament........................	2420,4580	0.193, 0.31
Tin, bright, tinned iron sheet................	76	0.043, 0.064
Tungsten:		
Filament, aged...........................	80, 6000	0.032, 0.35
Filament................................	6000	0.39
Zinc:		
Commercial, 99.1% pure, polished...........	440, 620	0.045, 0.053
Oxidized by heating at 750°F................	750	0.11
Galvanized sheet iron:		
Fairly bright...........................	82	0.228
Gray, oxidized..........................	75	0.276

B. Refractories, Building Materials, Paints, and Miscellaneous

Asbestos board...........................	74	0.96
Asbestos paper............................	100, 700	0.93, 0.945

Table 4-1. The Normal Total Emissivity of Various Surfaces.
(Continued)

Surfaces	°F	ε
Brick:		
Red, rough, but no gross irregularities	70	0.93
Silica, unglazed, rough	1832	0.80
Silica, glazed, rough	2012	0.85
Grog brick, glazed	2012	0.75
See Refractory materials, below		
Carbon:		
T-carbon, 0.9% ash	260, 1160	0.81, 0.79
Carbon filament	1900, 2560	0.526
Candle soot	206, 520	0.952
Lampblack:		
Water-glass coating	209, 362	0.959, 0.947
Water-glass coating	260, 440	0.957, 0.952
Thin layer on iron plate	69	0.927
Thick coat	68	0.967
0.003 in. or thicker	100, 700	0.945
Enamel, white, fused on iron	66	0.897
Glass, smooth	72	0.937
Gypsum, 0.02 in. thick or smooth on blackened plate	70	0.903
Marble, light gray, polished	72	0.931
Oak, planed	70	0.895
Oil layers on polished nickel (lubricating oil):		
Polished surface alone	68	0.045
+0.001 in. oil	68	0.27
+0.002 in. oil	68	0.46
+0.005 in. oil	68	0.72
+ ∞	68	0.82
Oil layers on aluminum foil (linseed oil):		
Aluminum foil	212	0.087
+1 coat oil	212	0.561
+2 coats oil	212	0.574
Paints, lacquers, varnishes:		
Snow-white enamel varnish on rough iron plate	73	0.906
Black shiny lacquer, sprayed on iron	76	0.875
Black shiny shellac on tinned iron sheet	70	0.821
Black-matte shellac	170, 295	0.91
Black lacquer	100, 200	0.80, 0.95
Flat black lacquer	100, 200	0.96, 0.98
White lacquer	100, 200	0.80, 0.95
Oil paints, 16 different, all colors	212	0.92, 0.96

TABLE 4-1. THE NORMAL TOTAL EMISSIVITY OF VARIOUS SURFACES.
(*Continued*)

Surfaces	°F	ϵ
Aluminum paints and lacquers:		
10% Al, 22% lacquer body, on rough or smooth surface	212	0.52
26% Al, 27% lacquer body, on rough or smooth surface	212	0.30
Other aluminum paints, varying age and Al content	212	0.27, 0.67
Aluminum lacquer, varnish binder, on rough plate	70	0.39
Aluminum paint, after heating to 620°F	300, 600	0.35
Paper, thin:		
Pasted on tinned iron plate	66	0.924
Pasted on rough iron plate	66	0.929
Pasted on black lacquered plate	66	0.944
Plaster, rough, lime	50, 190	0.91
Porcelain, glazed	72	0.924
Quartz, rough, fused	70	0.932
Refractory materials, 40 different	1110, 1830	
Poor radiators	....	0.65, 0.75
		0.70
Good radiators	....	0.80, 0.85
		0.85, 0.90
Roofing paper	69	0.91
Rubber:		
Hard, glossy plate	74	0.945
Soft, gray, rough (reclaimed)	76	0.859
Serpentine, polished	74	0.900
Water	32, 212	0.95, 0.963

considered may be used to give results that are sufficiently accurate for all practical purposes.

The following general conclusions may be drawn concerning the emissivity of surfaces:

1. Highly polished metals have low emissivities.

2. The emissivity of most substances increases with increase in temperature.

3. Most nonmetals have high emissivities.

4. The emissivity of any surface varies widely with the condition of that surface.

4-7. Radiation Heat Transfer between Surfaces Separated by Nonabsorbing Media. Equation (4-15) may be used to calculate the total emissive power of any body at any given temperature T. However, since all bodies at any temperature above absolute zero are continually radiating heat to their surroundings, even though they may at the same time be absorbing more heat than they emit, it is necessary to determine, for practical purposes, the net exchange of thermal radiation between a body and its surroundings. In this section, the radiation between the surfaces of solids separated by nonabsorbing mediums, such as air, nitrogen, oxygen, hydrogen, and chlorine, will be considered. Radiation between surfaces separated by gases capable of absorbing and emitting radiant energy will be considered later.

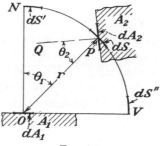

Fig. 4-5

The two factors involved in this problem are (1) shape and relative position of the radiating surfaces and (2) emissivity. It is convenient to treat the two factors separately by assuming temporarily that the surfaces being considered are perfect radiators and absorbers, i.e., black bodies, and that therefore only the factor of shape and relative position enters into the problem.

In Fig. 4-5 are shown cross sections of two bodies that for the purpose of this argument may be considered to be perpendicular to the plane of the paper. If the surface elements dA_1 and dA_2 at temperatures T_1 and T_2 are considered, the total emissive power of each may be written

$$E_1 = \sigma T_1^4 \qquad \text{Btu/(sq ft)(hr)} \qquad (4\text{-}16)$$
$$E_2 = \sigma T_2^4 \qquad \text{Btu/(sq ft)(hr)} \qquad (4\text{-}17)$$

and the rate of heat emission from each surface is

$$dq_1 = \sigma \, dA_1 \, T_1^4 \qquad \text{Btu/hr} \qquad (4\text{-}18)$$
$$dq_2 = \sigma \, dA_2 \, T_2^4 \qquad \text{Btu/hr} \qquad (4\text{-}19)$$

where $\sigma =$ Stefan-Boltzmann constant, equal to 0.173×10^{-8}. The element dA_1 will emit energy in all directions above it, but only a fraction of it will fall on the element dA_2. If this fraction is called f_2, and if the fraction of the energy being emitted by dA_2 that falls on dA_1 is called f_1, then the energy emitted by the surface dA_1 that falls on the surface dA_2 may be expressed by the equation

$$dq_{1-2} = \sigma f_2 \, dA_1 \, T_1{}^4 \qquad \text{Btu/hr} \qquad (4\text{-}20)$$

and, similarly,

$$dq_{2-1} = \sigma f_1 \, dA_2 \, T_2{}^4 \qquad \text{Btu/hr} \qquad (4\text{-}21)$$

In Fig. 4-5, r is the line joining the centers of the two surface elements dA_1 and dA_2 and ds is a surface element on a hemisphere of radius r, drawn with its center coinciding with the center of element dA_1. ON is perpendicular to dA_1, and PQ is perpendicular to dA_2. The element dA_1 will radiate energy to all parts of the surface of the hemisphere in varying degree. For example, an element ds', on the line ON which is perpendicular to dA_1 and ds', will receive more energy than an element ds'' on the line OV which lies practically on the surface dA_1. The decrease in the fraction of the energy leaving dA_1 and falling on any element ds on the sphere which occurs with an increase in the angle θ_1 can be shown to be in proportion to the cosine of the angle. Furthermore, the surface element ds receives energy in proportion to the ratio between its area and the area of the base of the hemisphere,* i.e., proportional to the ratio $ds/\pi r^2$. The element ds may be replaced by its equivalent $\cos \theta_2 \, dA_2$, since $\cos \theta_2$ equals ds/dA_2. The fraction f_2 may be written

$$f_2 = \frac{\cos \theta_1 \cos \theta_2 \, dA_2}{\pi r^2}$$

and by exactly similar reasoning

$$f_1 = \frac{\cos \theta_2 \cos \theta_1 \, dA_1}{\pi r^2}$$

Equations (4-20) and (4-21) may now be written

$$dq_{1-2} = \frac{\sigma \cos \theta_1 \cos \theta_2 \, dA_2 \, dA_1 \, T_1{}^4}{\pi r^2} \qquad (4\text{-}22)$$

$$dq_{2-1} = \frac{\sigma \cos \theta_1 \cos \theta_2 \, dA_2 \, dA_1 \, T_2{}^4}{\pi r^2} \qquad (4\text{-}23)$$

* See appendix for derivation.

or the net heat exchange between the two surface elements is

$$dq = dq_{1-2} - dq_{2-1} = \frac{\cos \theta_1 \cos \theta_2 \, dA_1 \, dA_2}{\pi r^2} \sigma(T_1^4 - T_2^4) \quad (4\text{-}24)$$

If $\cos \theta_1 \cos \theta_2 \, dA_2/\pi r^2$ is put equal to F_A, then the net heat exchange between the two surface elements is given by the expression

$$dq = \sigma F_A \, dA_1 \, (T_1^4 - T_2^4) \qquad \text{Btu/hr} \qquad (4\text{-}25)$$

F_A is usually called the configuration factor and is a function of θ_1, θ_2, A_2, and r. Values of F_A for a number of different practical cases, as determined by Hottel, are given in Table 4-2. It is fortunate that in some of the most important engineering problems the value of F_A is found to be unity.

Equation (4-25) applies only to "black" surfaces and must not be used for surfaces having emissivities very different from unity. Since most surfaces of industrial importance are not black, an expression for the radiant-heat exchange between such surfaces must contain the emissivity factor ϵ. To derive an expression that includes the emissivity it is necessary simply to introduce ϵ_1 and ϵ_2 in Eqs. (4-20) and (4-21). However, it is also necessary to remember that a surface that is not black will reflect part of the energy that falls upon it. Part of the reflected energy returns to the surface from which it originated, where it again undergoes partial reflection. This process goes on indefinitely; therefore the expressions for dq_{1-2} and dq_{2-1} become a series with an infinite number of terms. They can be evaluated quite simply,* however; and for the general case of two surfaces of area A_1 and A_2 at temperatures T_1 and T_2 having emissivities ϵ_1 and ϵ_2, the net heat exchange by radiation is

$$q = \sigma A_1 \left(\frac{\epsilon_1 T_1^4 f_2 - \epsilon_2 T_2^4 f_1 \dfrac{A_2}{A_1} + r_1 \epsilon_2 T_2^4 \dfrac{A_2}{A_1} f_1 f_2 - r_2 \epsilon_1 T_1^4 f_1 f_2}{1 - r_2 r_1 f_1 f_2} \right)$$
$$\text{Btu/hr} \quad (4\text{-}26)$$

where r_1 = reflectivity of the surface A_1, numerically equal to $1 - \epsilon_1$ if it is assumed that the two bodies follow Kirchhoff's law [Eq. (4-12)] exactly. To use Eq. (4-26) it is necessary to

* See appendix for derivation.

TABLE 4-2. RADIATION BETWEEN SOLIDS, FACTORS FOR USE IN
EQ. (4-27)

(*From Hottel*[7])

Surfaces between which radiation is being interchanged	Area A	F_A	F_e
1. Infinite parallel planes.	A_1 or A_2	1	$\dfrac{1}{\dfrac{1}{\epsilon_1} + \dfrac{1}{\epsilon_2} - 1}$
2. Completely enclosed body, small compared with enclosing body. (Subscript 1 refers to enclosed body.) See Ex. 4-1, p. 65.	A_1	1	ϵ_1
3. Completely enclosed body, large compared with enclosing body. (Subscript 1 refers to enclosed body.)	A_1	1	$\dfrac{1}{\dfrac{1}{\epsilon_1} + \dfrac{1}{\epsilon_2} - 1}$
4. Intermediate case between 2 and 3. (Incapable of exact treatment except for special shapes.) (Subscript 1 refers to enclosed body.)	A_1	1	$\epsilon_1 > F_e > \dfrac{1}{\dfrac{1}{\epsilon_1} + \dfrac{1}{\epsilon_2} - 1}$
5. Concentric spheres or infinite cylinders, special case of 4. (Subscript 1 refers to enclosed body.)	A_1	1	$\dfrac{1}{\dfrac{1}{\epsilon_1} + \dfrac{A_1}{A_2}\left(\dfrac{1}{\epsilon_2} - 1\right)}$*
6. Surface element dA and area A_2. There are various special cases of 6 with results presentable in graphical form. They follow as Cases 7, 8, 9.	dA	See special Cases 7, 8, 9†	$\epsilon_1 \epsilon_2$
7. Element dA and rectangular surface above and parallel to it, with one corner of rectangle contained in normal to dA.	dA	See Fig. 4-6	$\epsilon_1 \epsilon_2$
8. Element dA and any rectangular surface above and parallel to it. Split rectangle into 4 having common corner at normal to dA and treat as in Case 7.	dA	Sum of F_A's determined for each rectangle as in Case 7	$\epsilon_1 \epsilon_2$
9. Element dA and circular disk in plane parallel to plane of dA.	dA	Formula below‡	$\epsilon_1 \epsilon_2$
10. Two parallel and equal squares or disks of width or diameter D and distance between of L.	A_1 or A_2	Fig. 4-7, curves 1 and 2	$\epsilon_1 \epsilon_2$
11. Same as 10 except planes connected by nonconducting reradiating walls.	A_1 or A_2	Fig. 4-7, curve 3	$\epsilon_1 \epsilon_2$
12. Two equal rectangles in parallel planes directly opposite each other and distance L between.	A_1 or A_2	$\sqrt{F_A'F_A''}$§	$\epsilon_1 \epsilon_2$ or $\dfrac{1}{\dfrac{1}{\epsilon_1} + \dfrac{1}{\epsilon_2} - 1}$§
13. Two rectangles with common sides, in perpendicular planes.	A_1 or A_2	Fig. 4-8	$\epsilon_1 \epsilon_2$
14. Radiation from a plane to a tube bank (1 or 2 rows) above and parallel to the plane.	See Ex. 4-2, p. 65	Fig. 4-9	$\epsilon_1 \epsilon_2$

* This form results from assumption of completely diffuse reflection. If reflection is completely specular (mirrorlike), then $F_e = 1/[(1/\epsilon_1 + 1/\epsilon_2) - 1]$.

† A complete treatment of this subject, including formulas for special complicated cases and the description of a mechanical device for solving problems in radiation, is given by H. C. Hottel in *Mech. Eng.*, **52** (7), 699 (July, 1930).

‡ Case 9, R = radius of disk ÷ distance between planes; x = distance from dA to normal through center of disk ÷ distance between planes.

$$F_A = \frac{1}{2}\left[1 - \frac{x^2 + 1 - R^2}{\sqrt{x^4 + 2(1 - R^2)x^2 + (1 + R^2)^2}}\right]$$

§ F_A' = F_A for squares equivalent to short side of rectangle (Fig. 4-7, curve 2) and F_A'' = F_A for squares equivalent to long side of rectangle (Fig. 4-7, curve 2).
$F_e = \epsilon_1 \epsilon_2$ if the areas are small compared with L.
$F_e = 1/[(1/\epsilon_1 + 1/\epsilon_2) - 1]$ if the areas are large compared with L.

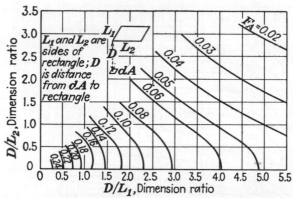

FIG. 4-6. Radiation between surface element and rectangle above and parallel to it.

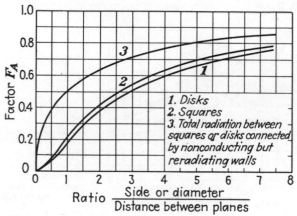

FIG. 4-7. Direct radiation between equal disks or squares in parallel planes directly opposed.

evaluate f_1 and f_2. However, for many practical purposes Eq. (4-26) may be replaced by an expression of the form

$$q = \sigma F_e F_A A_1 (T_1^4 - T_2^4) \qquad \text{Btu/hr} \qquad (4\text{-}27)$$

where F_e = factor to allow for the departure of the two surfaces from complete blackness, a function of the emissivities ϵ_1 and ϵ_2 as well as the configuration of the surfaces, and F_A = configuration factor, a function of the configuration of the surfaces. Table 4-2 presents values of F_e and F_A for various practical cases. Some of the values of F_e and F_A are derived directly from

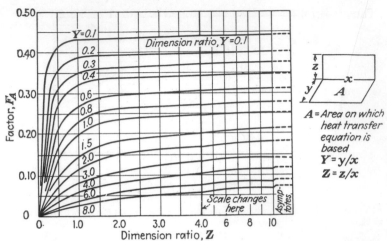

FIG. 4-8. Radiation between adjacent rectangles in perpendicular planes.

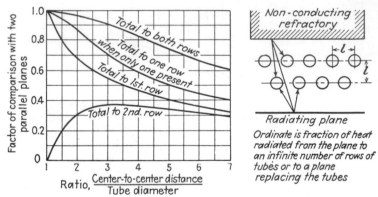

FIG. 4-9. Radiation from a plane to one or two rows of tubes above and parallel to the plane.

Eq. (4-26). For example, in the case of two infinite parallel planes where $A_1 = A_2$ and $f_1 = f_2 = 1$ it is seen that Eq. (4-26) becomes

$$q = A_1 \sigma (T_1{}^4 - T_2{}^4) \frac{\epsilon_1 \epsilon_2}{1 - r_1 r_2}$$

or since $r_1 = 1 - \epsilon_1$ and $r_2 = 1 - \epsilon_2$, then

$$q = A_1 \sigma (T_1{}^4 - T_2{}^4) \frac{1}{(1/\epsilon_1) + (1/\epsilon_2) - 1} \qquad \text{Btu/hr} \quad (4\text{-}28)$$

Thus it is found that for the case of infinite parallel planes

$$F_A = 1 \quad \text{and} \quad F_e = \frac{1}{(1/\epsilon_1) + (1/\epsilon_2) - 1}$$

This corresponds to the values shown in Table 4-2. Values of F_e and F_A for other cases have been worked out by various investigators and are included in Table 4-2.

It is possible with Table 4-2 and its accompanying graphs to treat many problems that may arise. For special surfaces that do not fall under any of the cases included in the table, the reader is referred to Hottel's original paper (see footnote† at end of Table 4-2). The following examples will serve to illustrate the use of Table 4-2 and the accompanying graphs.

Example 4-1. Determine the heat lost by radiation per foot length of 3-in. steel pipe at 500°F, if (a) located in a large room with red brick walls at a temperature of 80°F, and (b) enclosed in a 10- by 10-in. red brick conduit at a temperature of 80°F.

a. From Table 4-2, Case 2, it is seen that the heat transfer is independent of the area and emissivity of the enclosing surface (the walls) and is dependent only on A_1, the area of the pipe, which is $1 \times \pi \times 3.5 \div 12 = 0.916$ sq ft/ft of length, and ϵ_1 the emissivity of the oxidized steel pipe, which is 0.79 as shown by Table 4-1, page 55. The heat transfer by radiation is found by substituting these values in Eq. (4-27) in the following manner:

$$\begin{aligned} q &= \sigma F_e F_A A_1 (T_1^4 - T_2^4) \\ &= 0.173 \times 10^{-8} \times 0.79 \times 1 \times 0.916(960^4 - 540^4) \\ &= 957 \text{ Btu/(hr)(ft of pipe)} \end{aligned}$$

b. This case is closely approximated in Case 5 of Table 4-2. Here the heat transfer by radiation is dependent on the area of the conduit as well as the pipe area. The emissivity ϵ_2 of the brick conduit (from Table 4-1, page 57) is 0.93. The area A_1 of the pipe is 0.916 sq ft/ft, and the area A_2 of the conduit is $1 \times 40 \div 12 = 3.33$ sq ft/ft. As before, ϵ_1 is 0.79, and from Table 4-2

$$\begin{aligned} F_e &= \frac{1}{(1/\epsilon_1) + (A_1/A_2)[(1/\epsilon_2) - 1]} \\ &= \frac{1}{(1/0.79) + (0.916/3.33)[(1/0.93) - 1]} \\ &= 0.78 \end{aligned}$$

and

$$\begin{aligned} q &= 957 \times \frac{0.78}{0.79} \\ &= 945 \text{ Btu/(hr)(ft of pipe)} \end{aligned}$$

Example 4-2. A pulverized-fuel furnace has floor dimensions 16 by 20 ft. Ten feet above and approximately parallel to the floor is the first row of a nest of tubes which fill the furnace top. The floor, which is of refractory material, is at a temperature of 2240°F, and the tubes are at 540°F. The side walls (also refractory) are at an average temperature of 2240°F. The tubes are 3 in. o.d. and on 6-in. centers, staggered. Find (a) the radiant-heat transfer from the floor directly to the entire nest of tubes above; (b) the total net radiation to the first two rows of tubes from the floor and side walls if a baffle is placed just above the second row of tubes.

a. In this example Case 12, Table 4-2, applies.

$$A_1 = A_2 = 16 \times 20 = 320 \text{ sq ft}$$

From Fig. 4-7, $F_A' = 0.34$, since $D/L = {}^{16}\!/_{10} = 1.6$, and $F_A'' = 0.42$, since $D/L = {}^{20}\!/_{10} = 2$. F_A then equals $\sqrt{0.34 \times 0.42} = 0.378$. The emissivity ϵ_1 of the tube bank must be assumed higher than 0.79 (ϵ for oxidized steel, Table 4-1) since some of the radiation that escapes the first row of tubes will subsequently be absorbed by other tubes in the bank. For most practical purposes the emissivity ϵ_1 may be assumed equal to 0.9. The emissivity of the refractory floor, according to Table 4-1, is 0.70. Hence $F_e = \epsilon_1 \epsilon_2 = 0.9 \times 0.7 = 0.63$, and from Eq. (4-27)

$$q = 0.173 \times 10^{-8} \times 0.63 \times 0.378 \times 320(2700^4 - 1000^4)$$
$$= 6,876,000 \text{ Btu/hr}$$

b. The total radiation from the four walls to the tube bank, if there were no baffle present, is covered in Case 13. Referring to Fig. 4-8, for the end walls, $Y = {}^{20}\!/_{16} = 1.25$, and $Z = {}^{10}\!/_{16} = 0.625$; hence $F_A = 0.14$. For the side walls, $Y = {}^{16}\!/_{20} = 0.8$, and $Z = {}^{10}\!/_{20} = 0.5$; hence $F_A = 0.17$. For the four walls, $F_A = 2(0.14 + 0.17) = 0.62$. From part a, $F_e = 0.63$; hence the heat received from the four walls, based upon the area of the tube bank, is

$$q = 0.173 \times 10^{-8} \times 0.63 \times 0.62 \times 320(2700^4 - 1000^4)$$
$$= 11,278,000 \text{ Btu/hr}$$

The total from walls and floor is $6,876,000 + 11,278,000 = 18,154,000$ Btu/hr. The effect of the baffle is covered in Case 14 of Table 4-2. From Fig. 4-9, tube pitch $\div$ tube diameter $= {}^6\!/_3 = 2$, and the factor for comparison with two parallel planes is 0.97; hence, for part b

$$q = 18,154,000 \times 0.97$$
$$= 17,609,000 \text{ Btu/hr}$$

4-8. Solar, Celestial, and Terrestrial Radiation.

The problem of radiant-heat exchange between a body in the open and its surroundings is of importance in computing heating and cooling requirements of buildings, aircraft, storage tanks, and other devices which are exposed. The exact solution of such a problem is extremely complex and should be treated as a non-steady-state problem since the position of the sun and the temperatures of the surroundings vary with time. An approximate solution over a short period of time, however, may be obtained by treating it as a steady-state problem and applying the laws of radiant-heat exchange.

In the general case the surface of a body in the open will exchange radiant energy with the sun, the sky, and the earth. The net energy received by the surface may be expressed as

$$q = q_{SU} + q_{SK} + q_E - q_S \quad \text{Btu/hr} \quad (4\text{-}29)$$

where q_{SU} = absorbed portion of the energy received by the surface directly from the sun.

q_{SK} = absorbed portion of the energy received by the surface from the sky.

q_E = absorbed portion of the energy received by the surface from the earth.

q_S = total energy emitted from the surface.

The portion of the direct energy from the sun which is absorbed by a surface is

$$q_{SU} = \sigma \epsilon_{SU} t_r A F_A T_{SU}{}^4 \quad \text{Btu/hr} \quad (4\text{-}30)$$

where ϵ_{SU} = absorptivity of the surface for solar radiation (different from its emissivity; see Table 4-3).*

t_r = transmissivity of the earth's atmosphere.

A = surface area.

F_A = configuration factor.†

T_{SU} = absolute temperature of the sun (approximately 10,000°F abs).

The value of t_r, the transmissivity of the earth's atmosphere, is affected by the altitude of the surface receiving solar energy, the latitude, the time of year, time of day, and type of sky conditions. The configuration factor F_A is affected by the same factors with the exception of the sky conditions. If no atmosphere existed between the surface and the sun, the solar radiation falling on a surface normal to the sun's rays would be approximately 420

---

* Certain surfaces differ greatly from gray bodies; i.e., their emissivities are very different at widely different temperatures. For such surfaces there is no similarity between their ability to absorb high temperature (short-wavelength) radiation and their emissivities at ordinary temperatures.

† For flat surfaces, F_A is found from Eq. (4-24) to be $(r/R)^2 \cos \phi$ where r = radius of the sun, R = distance from the sun to the surface, and ϕ = angle between the sun's rays and the normal to the surface. See S. M. Marco, Effect of Solar Radiation on a Body at High Altitudes, "Nepa Heat Transfer Symposium," völ. 1, p. 230, Oak Ridge, Tenn., 1948.

Btu/(hr)(sq ft). Hence a flat surface normal to the sun's rays located in space outside the earth's atmosphere would receive solar radiation at this rate. This would be approximately true for any surface at altitudes higher than 50,000 ft. As the sun's rays pass through the earth's atmosphere some of the energy will be absorbed by the dust, water vapor, and carbon dioxide. In addition, some of the energy will be diffusely reflected, thus scattering it, and as a consequence only a fraction of the direct energy which would fall on a surface normal to the sun's rays outside the earth's atmosphere would reach the same surface if it were located inside the earth's atmosphere. As a consequence

TABLE 4-3. ABSORPTIVITIES OF VARIOUS SURFACES FOR
SOLAR RADIATION*

Substance	ϵ_{SU}
Building materials:	
Brick, red............................	0.70–0.77
Clay tiles, red and red-brown.........	0.65–0.74
Slate...............................	0.79–0.93
Other roofing materials:	
Galvanized iron, new..............	0.66
Galvanized iron, dirty.............	0.89
Roofing paper....................	0.88
Asphalt.........................	0.89
Paints:	
Black flat...........................	0.97–0.99
White flat...........................	0.12–0.26
Metals:	
Aluminum polished.................	0.26
Copper polished...................	0.26
Iron polished......................	0.45
Iron oxide (red)...................	0.74
Duralumin........................	0.53
Monel metal......................	0.43
Miscellaneous:	
White paper.......................	0.27
Asphalt pavement.................	0.85

* From M. Fishenden and O. A. Saunders, "Calculations of Heat Transmission," H. M. Stationery Office, London, 1932.

of these considerations, it is convenient to rewrite Eq. (4-30) in terms of values of direct solar radiation falling on a surface normal to the sun's rays which have been determined by measurement. Thus,

$$q_{SU} = \epsilon_{SU} H_{SU} A \cos \phi \qquad \text{Btu/hr} \qquad (4\text{-}31)$$

where ϕ = angle between the sun's rays and the normal to the surface, and $H_{SU} = (\sigma t_r T_{SU}{}^4 F_A)/\cos \phi$ is the direct solar radiation which would fall on the surface if it were normal to the sun's rays. Note that $A \cos \phi$ is the area of the surface projected onto a plane normal to the sun's rays. Values of H_{SU} as measured[4] are expressed in terms of the angle made by the sun and a line normal to the earth's surface (a vertical line). This angle, which is denoted by the symbol β_1, is the complement of the solar altitude (angle between the sun and a horizontal plane). The curve marked H_{SU} in Fig. 4-10 shows the way in which H_{SU} varies with β_1 for sea-level conditions and the values are representative of those which might be obtained on a clear summer day. The value of $\cos \phi$ of Eq. (4-31) may be determined by the equation

$$\cos \phi = \cos |\beta_1 - \beta_2| - \sin \beta_1 \sin \beta_2 (1 - \cos |\alpha_1 - \alpha_2|) \quad (4\text{-}32)$$

where β_1 = angle between the sun and the vertical.*

β_2 = angle between the normal to the surface and the vertical.

α_1 = azimuth angle of the sun.

α_2 = azimuth angle of the normal to the surface.

Values of β_1 and α_1 are plotted in Fig. 4-10 against the mean sun time for the hours from 6 A.M. to 6 P.M. for various northern latitudes. The curve of α_1 for 30 deg may be used for 25 and 35 deg latitude, and the curve shown for 45 deg may be used for 40 and 50 deg latitude. The local time should be corrected to the sun time on the basis of 1 hr for every 15 deg of longitude that the surface location is away from the standard meridian.

The symbol $|\beta_1 - \beta_2|$ indicates the absolute value of $(\beta_1 - \beta_2)$, and $|\alpha_1 - \alpha_2|$ is the absolute value of $(\alpha_1 - \alpha_2)$. The value of $|\beta_1 - \beta_2|$ must be less than 90 deg, while the value of $|\alpha_1 - \alpha_2|$ must be less than 90 deg or greater than 270 deg for the surface to *see* the sun. Values of these angles different from those indicated mean that the surface does not *see* the sun and therefore receives no direct solar radiation.

The value of q_{SK}, the portion of the energy received from the sky which is absorbed by the surface, may be expressed as

$$q_{SK} = \sigma \epsilon_{SK} A T_{SK}{}^4 \quad \text{Btu/hr} \quad (4\text{-}33)$$

where T_{SK} is the equivalent absolute temperature of the sky and

* See Fig. 4-11 for definitions of these angles.

ϵ_{SK} is the absorptivity of the surface for sky radiation. The value of T_{SK} is difficult to express in any simple fashion. For cold clear nights a value of 410°F abs (−50°F) is sometimes used. This value is based upon measurements of nighttime

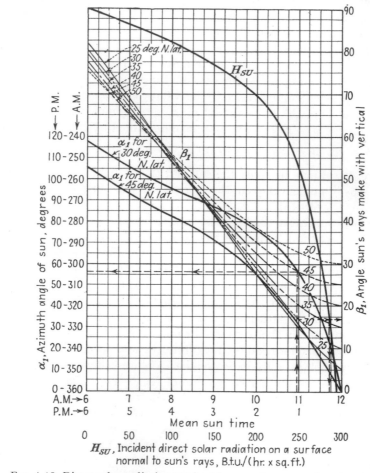

FIG. 4-10. Direct solar radiation received by a surface normal to the sun and solar angles for the period from May to August in northern latitudes.

radiation. For nighttime radiation the value of ϵ_{SK} may be assumed equal to the value of ϵ, the emissivity of the surface as given in Table 4-1. For the daytime, however, the value of T_{SK} is dependent not only on the temperature of the atmosphere and its transmissivity to outer-space radiation but it is also dependent upon the position of the sun relative to the surface. This is true because much of the sky radiation in the daytime is reflected solar radiation. For this reason it is more convenient to express the daytime sky radiation as a fraction of the solar radiation. Thus,

$$q_{SK} = f q_{SU} \qquad \text{Btu/(hr)(sq ft)} \qquad (4\text{-}34)$$

where f — ratio of sky radiation to direct solar radiation and is dependent upon β_1, the angle the sun makes with the vertical,

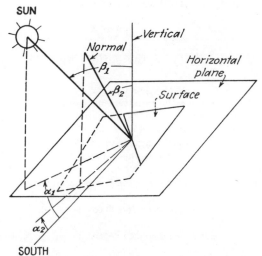

FIG. 4-11. Definitions of solar and surface angles.

and the condition of the sky. Values of f for various values of β_1 which may be used for very rough approximations are given in Table 4-4. These values apply to a horizontal surface on a

TABLE 4-4. APPROXIMATE RATIO OF SKY RADIATION TO DIRECT SOLAR RADIATION RECEIVED BY A HORIZONTAL SURFACE ON A CLEAR DAY

Vertical angle of sun β_1	Ratio f	Vertical angle of sun β_1	Ratio f
0	0.16	50	0.26
10	0.17	60	0.32
20	0.18	70	0.44
30	0.19	80	0.71
40	0.22	85	1.33

clear day.   Values of f for surfaces other than horizontal are only slightly smaller than those shown.

The term q_E in Eq. (4-29) may be determined by the equation

$$q_E = \sigma F_e F_A A T_E^4 \qquad (4\text{-}35)$$

where F_e and F_A = factors given in Table 4-2.

$\qquad T_E$ = absolute temperature of the earth's surface.

$\qquad A$ = surface area.

The total energy emitted from the surface is

$$q_S = \sigma \epsilon A T_S^4 \qquad (4\text{-}36)$$

where ϵ is the emissivity of the surface and T_S is its absolute temperature.

Example 4-3.   Determine the radiation heat exchange between a 15- by 25-ft new galvanized iron roof and its surroundings at 1:30 P.M. local time on a clear day.   The roof is situated at 30 deg north latitude and at a longitude where the local time is ½ hr ahead of the sun time.   It makes an angle of 30 deg with the horizontal, faces directly west, and has a temperature of 100°F.

The direct solar radiation from the sun is computed with Eq. (4-31) which is

$$q_{SU} = \epsilon_{SU} H_{SU} A \cos \phi$$

To find H_{SU} and $\cos \phi$ it is necessary to determine α_1, α_2, β_1, and β_2.   These values are:

$\alpha_1 = 56°$ (from Fig. 4-10 for 1 P.M. sun time)
$\alpha_2 = 90°$ (west direction)
$\beta_1 = 17°$ (from Fig. 4-10 for 1 P.M. sun time)
$\beta_2 = 30°$ (normal to surface makes an angle of 30 deg with the vertical)
$H_{SU} = 287$ Btu/(hr)(sq ft) (from Fig. 4-10 for $\beta_1 = 17°$)

$\cos \phi$ may now be determined from Eq. (4-32):

$$
\begin{aligned}
\cos \phi &= \cos |17 - 30| - \sin 17 \sin 30 \,(1 - \cos |56 - 90|) \\
&= \cos 13 - \sin 17 \sin 30 \,(1 - \cos 34) \\
&= 0.9744 - 0.2924 \times 0.500 \,(1 - 0.8290) \\
&= 0.9494
\end{aligned}
$$

and ϵ_{SU} from Table 4-3 is 0.66, hence

$$
\begin{aligned}
q_{SU} &= 0.66 \times 287 \times (15 \times 25) \times 0.9494 \\
&= 67{,}200 \text{ Btu/hr (direct solar radiation)}
\end{aligned}
$$

The radiation received from the sky is, according to Eq. (4-34),

$$q_{SK} = fq_{SU}$$

where $f = 0.18$ (from Table 4-4 for $\beta_1 = 17°$).
  $q_{SU} = 67,200$.
Thus

$$q_{SK} = 12,100 \text{ Btu/hr}$$

The radiation received from the earth may be determined to be negligibly small by observing that if the roof and the earth were considered to be black bodies, Eq. (4-22) could be integrated to determine the rate at which radiant energy from the earth would fall onto the roof.   Inspection of the geometry involved in this problem shows that both θ_1 and θ_2 are very large angles. The angle θ_1, which is the angle between the normal to an element dA_1 on the earth and the line joining the centers of dA_1 and dA_2, rapidly approaches 90 deg, hence cos θ_1 rapidly approaches zero.

The radiation emitted by the surface is computed from Eq. (4-36).   This is

$$q_S = \sigma \epsilon A T_s^4$$

where $\sigma = 0.173 \times 10^{-8}$.
  $\epsilon = 0.228$ (from Table 4-1).
  $A = 15 \times 25 = 375$ sq ft.
  $T_s = 460 + 100 = 560°F$ abs.
hence

$$q_S = 0.173 \times 10^{-8} \times 0.228 \times 375 \times 560^4$$
$$= 14,600 \text{ Btu/hr}$$

The net energy received by the surface, from Eq. (4-29), is

$$q = q_{SU} + q_{SK} + q_E - q_S$$
$$= 67,200 + 12,100 + 0 - 14,600$$
$$= 64,700 \text{ Btu/hr}$$

4-9. Coefficient of Radiant-heat Transfer.
It is sometimes desirable to evaluate the net exchange of radiation of a body small in size compared with the enclosure, e.g., a steam pipe in a room, by the simplified equation

$$q = h_r A (t_s - t_r) \qquad \text{Btu/hr} \qquad (4\text{-}37)$$

instead of by Eq. (4-27).   In Eq. (4-37),
  h_r = coefficient of radiant-heat transfer from solid to solid, expressed in Btu/(hr)(sq ft)(°F difference between enclosed and enclosing surfaces).
  A = surface of the enclosed body, sq ft.
  t_s = its temperature, °F.
  t_r = temperature of the enclosing surface, °F.

If Eqs. (4-37) and (4-27) are combined, the value of h_r is found to be

$$h_r = \frac{\sigma F_e F_A (T_s{}^4 - T_r{}^4)}{t_s - t_r} \qquad \text{Btu/(hr)(sq ft)(°F)} \qquad (4\text{-}38)$$

A knowledge of the value of h_r is essential in determining the surface coefficient h for heat transfer from a surface by both convection and radiation, since the surface coefficient h is the sum of the convection coefficient h_c and the radiant-heat-transfer coefficient h_r.

4-10. Heat Radiation by Gases. The theory of radiation from gases is beyond the scope of this text. It is an involved field of study in which the literature is beginning to be put into a usable form. For a more complete treatment of this subject the reader is referred to more extensive works on heat transfer.[5] For the purpose of this text it will be sufficient to give the more important formulas and data that may be used for solving problems involving this type of radiation.

TABLE 4-5. AVERAGE LENGTHS OF RADIANT BEAMS IN VARIOUS
GAS SHAPES
(*From Hottel[5]*)

Shape	L
1. Sphere	$\frac{2}{3}$ × diameter
2. Infinite cylinder	1 × diameter
3. Space between infinite parallel planes	1.8 × distance between planes
4. Cube	$\frac{2}{3}$ × side
5. Space outside infinite bank of tubes with centers on equilateral triangles; tube diameter equals clearance	2.8 × clearance
6. Same as (5) except tube diameter equals one-half clearance	3.8 × clearance

Gases do not emit radiation at all wavelengths as do solids, but they may emit and absorb radiation in several bands of wavelengths. Of the gases encountered in heat-transfer equipment, carbon monoxide, the hydrocarbons, water vapor, carbon dioxide, sulfur dioxide, ammonia, and hydrogen chloride are the only ones that absorb or emit radiation over a sufficiently wide band of wavelengths to warrant consideration in engineering calculations. Hottel[6] states that the energy emitted from a gas mass to a unit area of its bounding surface is a function of the gas temperature t_g and the product term PL, where P is the partial pressure, in atmospheres of the radiating gas, such as carbon dioxide or water vapor, and L is the average length in feet of the path of the radiant beams to the surface in question. L is dependent upon the shape of the gas mass. Values of L for different shapes are given in Table 4-5.

In general, the net radiant-heat interchange between a gas and a solid surface may be found from an equation of the type

$$q = A\epsilon(I_G - I_S) \qquad \text{Btu/hr} \qquad (4\text{-}39)$$

where A = area of solid surface.

ϵ = emissivity of solid surface.

I_G = intensity factor which depends on the product term PL and the temperature of the gas.

I_S = intensity factor which depends upon the product term PL and the temperature of the solid surface.

Values of I_G and I_S for carbon dioxide, water vapor, and mixtures of water vapor and carbon dioxide are shown graphically in Figs. A-12 to A-14.

4-11. Radiation from Flames. Flames may be divided, for the purpose of convenience, into two general classes, viz., nonluminous flames and luminous flames. Flames that are nonluminous are normally invisible in the furnace; or if they are visible, they appear as bluish flames which are obtained where combustion is very rapid. Gases high in methane normally burn without luminosity if there is a high degree of primary aeration. Luminous flames are those made luminous by incandescent solid particles such as soot or powdered coal suspended in the flame. Nonluminous flames, like gases, are selective radiators; i.e., they emit radiation in only certain bands of the spectrum and not at all wavelengths. Hence radiation formulas for nonluminous flames are the same as those for gases.

Luminous radiation follows more nearly the Stefan-Boltzmann law of radiation that applies to solids. A luminous flame is different from a solid in that it is partly transparent, and therefore the radiation from it is dependent on the concentration of the particles in the flame as well as the fourth power of the absolute temperature. If the luminosity is due to solid particles caused by the combustion, i.e., soot, then the equation for net radiation exchange between a flame and an enclosing wall is given by a formula similar to Eq. (4-27) for a small body completely enclosed. This equation is

$$q = \sigma A (T_F{}^4 - T_W{}^4) \epsilon_F \epsilon_W \qquad \text{Btu/hr} \qquad (4\text{-}40)$$

where A = area of the flame envelope, sq ft.

T_F = absolute temperature of the flame, °F abs.

T_W = absolute temperature of the wall, °F abs.

ϵ_W = emissivity of the wall.

ϵ_F = emissivity of the flame as given in Figs. A-15 and A-16.

If the luminosity is due to powdered coal, the same formula may be used, except that in this case ϵ_F, the emissivity of the flame, is given by the expression

$$\epsilon_F = 1 - e^{-x} \qquad (4\text{-}41)$$

where

$$x = \frac{238L[(1 - V)\rho_0/\rho_1]^{2/3}}{g_B \rho_0 d_0 T_F} \qquad (4\text{-}42)$$

In this expression

L = effective flame thickness as given in Table 4-5.

V = fraction of volatile matter plus moisture in the coal as fired.

ρ_0 and ρ_1 = initial densities of the coal and of the coked coal, respectively.

g_B = pounds of combustion products per pound of coal.

d_0 = initial average particle diameter, in.

T_F = average temperature of the flame, °F abs.

To eliminate the necessity of evaluating all these terms, Fig. A-17 may be used.

Problems

1. Heat radiates from a small sphere suspended in a large room. The surface of the sphere of 1.5 sq ft is at a temperature of 580°F, and the inside surfaces of the room are at a temperature of 80°F. Assuming the emissivity of the sphere surface to be 0.30, calculate the radiant-heat transfer in Btu/hr.

2. A 12- by 12-in. ingot iron casting 60 in. high is stripped from its mold while its temperature is 1740°F. If it stands on end on the floor of a large foundry having wall, floor, and roof temperatures of 80°F, what is the rate of radiant-heat exchange between the casting and the roof, wall, and floor surfaces?

3. Determine the net radiant-heat exchange between two parallel oxidized iron plates when the distance between them is 2 ft and each plate measures 10 by 10 ft. The surface of one is at 200°F and of the other is at 100°F.

4. If the distance between the plates of Prob. 3 is reduced to 1 in., what percentage increase in net radiant-heat exchange will result?

5. A steam radiator with enveloping radiating surface 60 in. long by 26 in. high by 12 in. deep is supported above the floor of a large room. If the radiator surface, painted with lacquer containing 10 per cent aluminum, is at a temperature of 200°F and the room surfaces are at 70°F, what is the rate of radiant-heat exchange between the radiator and the surfaces of the room?

6. A 6-in. bare oxidized steel pipe (6.625 in. o.d.) carries steam at a temperature of 540°F and is located in a large room having wall temperatures of 80°F. Determine the net radiant-heat exchange in Btu/(hr)(ft) between the pipe and the room, (a) if the surface temperature of the pipe is assumed to be the same as the steam temperature; (b) if the pipe is covered with a single layer of asbestos paper and the temperature of the outer surface of the paper is 530°F; (c) if the asbestos paper is painted with aluminum paint containing 26 per cent aluminum and the surface temperature is 530°F.

7. Determine the net radiant-heat exchange between the closed ends of a hollow cylinder 12 in. in diameter and 4 in. long. The ends are of Cr-Ni alloy with one end heated electrically to a temperature of 1900°F and the other end maintained at 125°F. The side wall of the cylinder is substantially nonconducting and reradiating.

8. A horizontal metal plate which is completely insulated except for one exposed surface of 100 sq ft is heated electrically to maintain a temperature of 200°F on the exposed surface. Heat is transferred from this surface by both radiation and conduction through air (assumed to be without motion) to a second plate of similar size maintained at a surface temperature of 100°F. The distance between the parallel plates is 0.25 in., and each plate has an emissivity of 0.20. What is the watts input to the warmer plate?

9. A 6-in. bare oxidized steel pipe (6.625 in. o.d.) carrying steam at a high temperature is enclosed in a 12- by 12-in. red-brick conduit having wall temperatures somewhat lower than the pipe temperature. What percentage of the heat lost by radiation from the pipe to the walls of the conduit would be saved if the conduit walls were painted with aluminum paint containing 26 per cent aluminum?

10. A 12- by 24-in. open pan containing water at 140°F is placed in the exact center of a 16- by 20-ft room having a ceiling temperature of 75°F. If the distance from the water surface to the ceiling is 10 ft and the ceiling is painted with an oil paint, what is the net radiant-heat exchange between the water surface and the ceiling?

11. A 12- by 12-in. fire door in a brick kiln is at a distance of 10 ft from a wall 30 ft wide and 15 ft high. The door is so located that a normal from the center of the door strikes the wall midway from the ends and 3 ft above the floor. The opening into the kiln may be considered a black body; the wall is painted with flat oil paint. The temperature inside the kiln is 2540°F, and the wall temperature is 140°F. What is the net radiant-heat exchange between the opening and the wall when the fire door is open?

12. A 30- by 30-ft room, 10 ft high, is heated by hot-water pipes laid beneath an oak floor. The walls and ceiling of the room are coated with an oil paint. The surface temperature of the floor is 80°F, and the surface temperature of the ceiling is 60°F. What is the rate of radiant-heat exchange between the floor and the ceiling? Assume the side walls to be nonconducting but reradiating.

13. For the room described in Prob. 12, what is the rate of radiant-heat exchange between the floor and the four walls when the surface temperature of the floor is 80°F and the average surface temperature of the walls is 60°F?

14. A room in a basementless house is heated by means of a panel-heating installation located in the ceiling. The ceiling is painted with an oil paint and the floor is covered with hard glossy-plate rubber tile. Room dimensions are 20 by 15 ft with a 10-ft ceiling height. Neglecting any effect due to the walls, determine the floor temperature for a radiant-heat loss from the ceiling to the floor of 4250 Btu/hr if the surface temperature of the ceiling is 115°F.

15. A well-insulated inside room, 20 ft long and 20 ft wide with a 10-ft ceiling height, is to be heated by means of a ceiling-panel installation. It is desired to maintain the surface of the floor at a temperature of 80°F. Determine the necessary ceiling-surface temperature to meet this requirement if the floor is to be of planed oak, the ceiling is to be painted with an oil paint, and it is estimated that the portion of the heating requirement to be supplied by radiation is 5000 Btu/hr. Assume the walls to be nonconducting but reradiating.

16. A corridor 10 ft wide and 40 ft long is heated by a baseboard radiator 12 in. high extending the full width of the 10-ft end. The surface of the radiator is covered with flat black lacquer, and the floor is of marble. Determine the net radiant-heat exchange between the radiator and the floor when the temperature of the radiator surface is 200°F and the average temperature of the floor surface is 80°F.

17. The net radiant-heat exchange between a 20- by 20-ft rough-plaster panel-heated ceiling and a 20- by 10-ft side wall, which has an emissivity of 0.80, is 2400 Btu/hr. If the surface temperature of the ceiling is 120°F, determine the average surface temperature of the side wall.

18. A 15- by 20-ft roof, covered with asphalt, absorbs solar radiation at 1:30 P.M. local time on a clear summer day. The roof is situated at 45 deg north latitude at a longitude 7.5 deg east of the standard meridian for its time zone and slopes directly south at an angle of 20 deg with the horizontal. Determine q_{SK}, the absorbed portion of the radiation received by the roof from the sky.

19. A 30- by 30-ft flat roof is covered with new galvanized iron. The net radiant-heat exchange between the roof and the atmosphere and the sun is 148,000 Btu/hr gained by the roof. If this occurs at noon (sun time) at a 40-deg north latitude on a clear summer day, what is the temperature of the roof surface?

20. A southwest wall of a building has dimensions of 30 by 12 ft. It is made of red brick and has a surface temperature of 100°F. The building is located at 45 deg north latitude and at a longitude of 3.75 deg west of the standard meridian for its time zone. The wall in question cannot *see* other buildings. Determine the rate of net radiant-heat transfer to the wall at 4 P.M. local time on a clear summer day. Assume the earth temperature to be 80°F and its emissivity to be 0.90. $F_A = 0.50$.

REFERENCES

1. Pringsheim, E.: *Verhandel. deut. physik. Ges.*, 1, 23, 215 (1899), and 2, 163 (1900).
2. Reiche, F.: "The Quantum Theory," p. 15, English translation by Hatfield and Brose, E. P. Dutton & Co., Inc., New York.
3. Planck, M.: *Verhandel. deut. physik. Ges.*, 2, 237 (1900).
4. Moon, P.: Proposed Standard Solar Radiation Curves for Engineering Use, *J. Franklin Inst.*, 230(5), 583–617 (1940).
5. McAdams, W. H.: "Heat Transmission," 3d ed., McGraw-Hill Book Company, Inc., New York, 1954.
6. Hottel, H. C.: *Trans. Am. Inst. Chem. Engrs.*, 19, 173 (1927).
7. Hottel, H. C.: Radiant Heat Transmission, *Mech. Eng.*, 52(7), 699 (1930).

CHAPTER 5

FUNDAMENTAL UNITS
AND DIMENSIONAL ANALYSIS

5-1. General. In the preceding chapters, use has been made of the familiar units of time, length, area, temperature, and heat. In the following chapters, which deal with the flow of fluids and with heat transfer by convection, not only these but a number of additional units will be used for describing the physical properties of fluids and the manner in which heat is transferred. The relationships of some of these units will first be reviewed, and then the principles of dimensional analysis will be introduced as an aid to the understanding of material which follows.

5-2. Force and Mass Units. Among the properties of fluids which have to be dealt with in a study of fluid flow and heat transfer are a number which involve units of either force or mass. One of these is the weight per unit volume. Weight per unit volume is commonly expressed in pounds per cubic foot and is termed *specific weight*. It is usually denoted by the symbol γ. The *density* of a substance, usually denoted by the symbol ρ, also is commonly expressed in pounds per cubic foot. The two terms, specific weight and density, are not synonymous, however, since in more precise language, specific weight is in units of pounds force per cubic foot, whereas density is in units of pounds mass per cubic foot. The proper application of one or the other of these terms can readily be made in calculations of fluid flow and heat transfer when it is established whether the problem deals essentially with the weight of the fluid or with its mass. In many other instances, however, confusion in the use of units of force and mass may occur unless differences in definitions of the units are noted and different concepts of weight are observed.

Whenever a body is weighed on a spring scale, the weight is a measure of the force of gravity exerted on the body; and that force is not necessarily the same at different locations. If, on the other hand, a body is weighed by balancing it against a known mass on a beam scale, the pull of gravity is exerted equally on both masses and the beam-balance comparison remains the same at all locations.

In most branches of science and engineering the prevailing concept of weight is the weight indicated by the spring scale—therefore expressed in units of force and not of mass. In many instances, however, it is desirable to establish the mass of a body from knowledge of its weight at any certain location where the acceleration due to gravity is known. The weight, expressed in units of force, can readily be converted into units of mass, provided the relationship of force and mass units is properly understood. Discussion of this relationship will follow, after the units of force and mass are defined.

Mass is a measure of the amount of matter. The pound mass is an amount, fixed, in this country, by act of Congress by reference to a standard pound body carefully preserved at the U.S. Bureau of Standards. It is the equivalent of 453.592,427,7 grams mass.

The pound force and the pound mass, as commonly defined and as used in this text, are both fixed values. The pound force is the force required to support the standard pound mass against gravity, *in vacuo*, in the standard locality—any locality in which the acceleration due to gravity equals 32.1739 ft/sec². Otherwise stated, it is the force which, if applied to the standard pound mass, supposed free to move, would give that mass the standard acceleration of 32.1739 ft/sec².

The pound force may likewise be defined as the force which, if applied to the standard slug of mass, supposed free to move, would give that mass an acceleration of 1 ft/sec². Use of the slug as the standard unit of mass is now quite common in many branches of science and engineering, whereas formerly the pound was the unit of mass generally used.

Comparison of the definitions of the pound force in terms of the pound mass and in terms of the slug shows that a slug equals 32.1739 pounds mass.

5-3. Relationship of Force and Mass. Newton stated that force F is proportional to mass m times acceleration a—*not equal* to mass times acceleration. In other words, F/ma is a constant with any system of units. Under all conditions, then,

$$F/ma = F_s/m_s a_s$$

where the subscript s denotes conditions established as a standard. Therefore,

$$F = \frac{F_s ma}{m_s a_s}$$

A preferred form for this equation is

$$F = \frac{1}{m_s a_s/F_s}\, ma \qquad (5\text{-}1)$$

since $m_s a_s/F_s$ defines the relationship that for each standard unit of force a standard mass is accelerated at a standard rate. In terms of the standard pound force lb_f, the standard pound mass lb_m, and the standard acceleration of 32.1739 ft/sec^2,

$$\frac{m_s a_s}{F_s} = \frac{lb_m}{lb_f} \times 32.1739 \text{ ft/sec}^2$$

and Eq. (5-1) may be written

$$F = \frac{1}{32.1739\ (lb_m/lb_f)\text{ft/sec}^2}\, ma \qquad (5\text{-}2)$$

The factor 32.1739 (lb_m/lb_f)ft/sec^2 is commonly denoted by g_c and is called a dimensional constant, or a constant of proportionality. When this substitution is made, Eq. (5-1) becomes

$$F = \frac{1}{g_c}\, ma \qquad (5\text{-}3)$$

From a comparison of Eq. (5-3) with Eq. (5-1) it should be noted that g_c equals $m_s a_s/F_s$, regardless of the units in which the standard mass is expressed. If m_s is the standard slug, which is the mass that is accelerated 1 ft/sec^2 by application of the standard pound force, then

$$g_c = \frac{\text{slugs} \times 1 \text{ ft/sec}^2}{lb_f} \qquad (5\text{-}4)$$

The foregoing examination of Newton's law shows that *units of mass may be converted into equivalent units of force by dividing by g_c.* Conversely, *units of force may be converted into equivalent units of mass by multiplying by g_c.* If force and mass are both expressed in pounds, g_c is 32.1739(lb_m/lb_f)ft/sec^2. If force is expressed in pounds and mass is in slugs, then g_c must be (slugs/lb$_f$)ft/sec^2 and must be numerically equal to unity. Applications of the foregoing procedure will now be made to show the relationship of terms which are commonly used to denote such physical properties of fluids as weight and mass, specific weight and density, and viscosity.

5-4. Weight. The magnitude of a force exerted on a body is commonly measured by the pull on a spring scale. As already observed, if the force is due to the gravitational effect of the earth, it indicates the *weight* of the body. The weight of a pound mass may or may not be one pound force, depending on the locality and the corresponding deviation of the pull of the earth from the standard which defines the pound force.

If in Eq. (5-3) the acceleration a is the acceleration due to gravity g, and m is the mass in pounds lb$_m$, then the force F represents the weight W in pounds lb$_f$ and the equation for weight becomes

$$W = \frac{g}{g_c} \text{lb}_m \qquad (5\text{-}5)$$

Example 5-1. If the value of g decreases 0.003 ft/sec^2 per 1,000 ft of elevation, determine the weight of a 100 pound mass at an elevation of 1 mile.

Solution

$$g = 32.1739 - (0.003 \times 5,280/1,000)$$
$$= 32.1739 - 0.01584 = 32.15806$$

$$W = \frac{g}{g_c} \text{lb}_m$$

$$= \frac{32.15806}{32.1739} \times 100 = 99.9508 \text{ lb}$$

The variation of g with latitude and elevation is so small that in many engineering calculations it is neglected. In this case, g and g_c have the same numerical value of 32.1739, and the weight of a pound mass is one pound force. It should be observed, however, that g and g_c are expressed in different units; they have different meanings, g representing the acceleration due to gravity, and g_c defining the relationship of the pound force to the pound mass and the standard acceleration. The factor g_c is used in various expressions solely for the purpose of avoiding confusion between the units of force or weight and mass.

5-5. Specific Weight and Density. Such terms as specific weight γ and density ρ, when both are in terms of pounds per cubic foot, are liable to be confused unless their relationship is observed by an expression in the form of Eq. (5-5). Dividing both sides of that equation by cubic feet shows that

$$\gamma = \frac{g}{g_c} \rho \tag{5-6}$$

Confusion between γ and ρ may produce no arithmetical error because of the similarity of the numerical values of g and g_c; but confusion in the units in which various terms are expressed will surely result if the relationship of γ and ρ is not observed.

5-6. Viscosity. Viscosity, the property by which a fluid offers resistance to shear, is obviously related to force rather than to mass, and yet the numerical values of the viscosity of fluids are published in a variety of units* of mass as well as of force— and not without purpose. In expressions which involve viscosity along with such a property as specific weight, since specific weight is expressed in terms of force, then viscosity should be expressed in terms of force. If, on the other hand, an expression involves viscosity and density, then viscosity should be expressed in terms of mass. If viscosity in terms of pounds force is μ_f and in terms of pounds mass is μ_m,

then $$\mu_f = \frac{\mu_m}{g_c} \tag{5-7}$$

Conversion factors for viscosity units are shown in the footnote.*

Example 5-2. The viscosity μ_m of water at 70°F has been expressed as 0.00066 $lb_m/(ft)(sec)$. Express this viscosity in terms of feet, seconds, and pounds force lb_f.

Solution

$$\mu_f = \frac{\mu_m}{g_c} = \frac{0.00066 \ lb_m/(ft)(sec)}{32.1739 \ (lb_m/lb_f)(ft)/(sec^2)}$$

$$= 0.000,020,5 \ \frac{(lb_f)(sec)}{sq \ ft}$$

---

* The relationships between the most common units of absolute viscosity are as follows:

lb_f-sec/sq ft	× 32.17	= $lb_m/(ft)(sec)$
lb_f-sec/sq ft	× 115,812	= $lb_m/(ft)(hr)$
$lb_m/(ft)(sec)$	× 3,600	= $lb_m/(ft)(hr)$
poises†	× 0.0672	= $lb_m/(ft)(sec)$
centipoises	× 0.000672	= $lb_m/(ft)(sec)$
centipoises	× 0.000,020,9	= lb_f-sec/sq ft
centipoises	× 2.42	= $lb_m/(ft)(hr)$

† 1 poise = 1 dyne sec/sq cm, or 1 gram (mass)/cm sec.

Viscosity is often expressed in pounds (force) times seconds (or hours) per square foot; or in the cgs system by the poise* or centipoise, where the poise is defined as 1 dyne sec/sq cm. Otherwise, viscosity may be expressed in units of mass as pounds (mass) or slugs per foot-second or per foot-hour; or in the cgs system as grams (mass) per centimeter-second. Viscosity, as expressed above, is often termed *absolute viscosity* in order to distinguish it from arbitrary viscosity units applied to commercial viscometers. It is also often termed *dynamic viscosity* to distinguish it from *kinematic viscosity*, which is the ratio of dynamic, or absolute, viscosity to the density of the fluid. The usual units of kinematic viscosity are feet squared per second or feet squared per hour, therefore involving only the kinematic quality of a function of length and time, in contrast to the dynamic quality of a function of force, length, and time, all three of which are present in the units of dynamic viscosity. In the cgs system, kinematic viscosity has the units of centimeters squared per second, called the stoke.†

5-7. Dimensional Analysis. The principles of dimensional analysis are very useful in correlating experimental data. There is nothing new or mysterious about analyzing experimental data from a dimensional viewpoint. Every student of physics or mechanics applies the methods of dimensional analysis to various simple problems in these fields. Many early physicists used these principles, but it is only recently that the practical value of dimensional analysis in the study of new problems has begun to be realized by the engineer.

The primary purpose of dimensional analysis is to correlate the various measurable physical quantities, such as length, time, and force, associated with a given phenomenon into a mathematical equation expressing the relationship among these quantities. It is not a complete analysis in the sense that differential equations of motion of a mechanical system are complete; but on the other hand, it is much more rapid and involves much less work than a detailed analysis by differential equations.

The results obtained by applying dimensional analysis, like any other method of analysis, are limited by the completeness and

---

* Named in honor of Jean Louis Poiseuille, who as early as 1840 stated the relationship of pressure drop to velocity for streamline flow in a tube.

† Named in honor of Sir G. G. Stokes, who was the first to use this unit.

correctness of the original assumptions. If all the variables involved in a problem are recognized, it is possible, by the methods of dimensional analysis, to predict by means of mathematical expressions the relationship among the variables before resorting to experimentation. In this respect, dimensional analysis is a very valuable tool, because it points out the manner in which any of the factors vary when the other quantities are varied and therefore enables the experimenter to determine which factors will be varied and which held constant in his experiment.

Usually it is desired to determine the dependence of one physical quantity upon a number of other physical quantities that are supposed to enter into a problem or experiment. It will be noted that dimensional analysis does not answer the question, "On what quantities does the result depend?" but rather "How does the result depend upon certain quantities which are known to affect it?" Furthermore, dimensional analysis does not replace experimentation, since it does not yield the numerical value of the constants involved that are necessary for the complete determination of the quantity sought. These constants can be determined only by direct experiment.

The methods of dimensional analysis depend upon a few fundamental rules which will be stated here without rigid proof.[2] The underlying rule in dimensional analysis is that the *dimensional formula* of every measured quantity is expressible as the product of powers of the *fundamental quantities* upon which it depends. This statement contains a number of concepts that are not well understood by most students. What is meant by a *dimensional formula* and what is meant by a *fundamental quantity* should be clearly understood before any profitable use can be made of the methods of dimensional analysis.

The *dimensions* of a physical quantity may be defined as the properties, independent of the magnitude of units used, by which a physical quantity may be described. In other words, *dimensions* are concerned with the *quality* of a physical quantity and not its magnitude. For example, the distance between two points on a bar may be described as 5 ft, or 60 in., or 152.4 cm. These are all different ways of describing the same thing, viz., the distance between two points, but they all have the common quality of being a length and not, for example, a volume or an area. The dimensional formula of this physical quantity is called length and is denoted by the symbol L. Similarly the dimensional formula of time is T. Length and time are physical quantities that can be measured easily and are usually considered as *fundamental quantities*, or *fundamental dimensions*, in any dimensional system. Other physical quantities are not always so directly measurable, and consequently their dimensional formulas are usually more complex. For example, in terms of length as a fundamental dimension the dimensional formulas of area and volume are, respectively, (length)^2 and (length)^3. It is customary to write this in symbols as follows: $A = L^2$ and $V = L^3$. These equations are read: "The dimensional formula of area is length squared, and the dimensional formula of volume is length cubed." The quantities area and volume are, in this case, said to be *secondary quantities*. In the example just used, area and volume are called secondary quantities since they are expressed in terms of the powers of the fundamental dimension L.

It is frequently possible to use a quantity as a fundamental dimension in one system of dimensions and as a secondary quantity in another system. This is seen to be the case in a dimension system in which, for example, mass is called a fundamental dimension and force is called a secondary quantity, which when expressed in terms of mass, length, and time is $F = MLT^{-2}$. In another system it might be more convenient to call force a fundamental dimension, thus expressing mass as a secondary quantity in terms of force, length, and time as $M = FL^{-1}T^2$. In any problem in which n physical quantities are involved and r of them are considered fundamental dimensions, then the remaining $n - r$ quantities are secondary quantities and must be expressible as products of powers of the fundamental dimensions. In any dimension system there should be just enough fundamental dimensions to express all the desired secondary quantities in terms of fundamental dimensions. There are many dimension systems in use in the various fields of science, but for the purpose of this text the more common engineering system is used. In this system the fundamental dimensions are force F, mass M, length L, time T, temperature θ, and heat H. Table 5-1 shows some of the more frequently used dimensional formulas of this system.

TABLE 5-1. DIMENSIONAL FORMULAS OF MECHANICAL AND
THERMAL QUANTITIES

Quantity	Symbol	Units in ft, lb, hr, Btu, and °F	Dimensional formulas
Acceleration of gravity......	g	4.18×10^8 ft/hr²	L/T^2
Angular velocity............	ω	1/hr	$1/T$
Area.....................	A	sq ft	L^2
Coefficient of heat transmission, film................	h	Btu/(hr)(sq ft)(°F)	$H/TL^2\theta$
Coefficient of thermal conductivity..................	k	Btu/(hr)(sq ft)(°F/ft)	$H/TL\theta$
Density..................	ρ	lb_m/cu ft	M/L^3
Dimensional constant........	g_c	4.18×10^8 (lb_m/lb_f)ft/hr²	ML/FT^2
Force*.....................	F	lb_f	F
Frequency.................	ν	1/hr	$1/T$
Heat.....................	q	Btu	H
Heat.....................	q	ft-lb_f	FL
Length....................	L, D	ft	L
Mass*....................	M	lb_m	M
Mechanical equivalent of heat	J	778 ft-lb_f/Btu	FL/H
Power.....................	P	ft-lb_f/hr	FL/T
Pressure..................	P	lb_f/sq ft	F/L^2
Rate of heat transfer per unit area.....................	q/A	Btu/(hr)(sq ft)	H/TL^2
Specific heat..............	c	Btu/(lb_m)(°F)	$H/M\theta$
Specific weight.............	γ	lb_f/cu ft	F/L^3
Stress....................	S	lb_f/sq ft	F/L^2
Temperature...............	t	°F	θ
Time.....................	T	hr	T
Velocity..................	V, u	ft/hr	L/T
Viscosity (absolute).........	μ_f	lb_f-hr/sq ft	FT/L^2
Weight*..................	W	lb_f	F

* Those quantities which involve mass may be converted to force dimensions by substituting FT^2/L for M. Similarly, those quantities which involve force may be converted to mass dimensions by substituting ML/T^2 for F.

A second important rule of dimensional analysis is known as the π theorem,* which states that any complete homogeneous equation† expressing the relationship between n measurable quantities and dimensional constants‡ such as $(\alpha, \beta, \gamma \ldots)$ in the form $f(\alpha, \beta, \gamma \ldots) = 0$ has a solution of the form

$$\phi(\pi_1, \pi_2, \pi_3, \ldots, \pi_{n-r}) = 0 \qquad (5\text{-}8)$$

where the number of π terms is $n - r$ independent products of the terms $\alpha, \beta, \gamma \ldots$, which are dimensionless in the fundamental units. In Eq. (5-8), n is the number of physical quantities or dimensional constants involved and r is the number of fundamental dimensions required to express them. Thus for five physical quantities involved, if they are expressed in terms of three fundamental dimensions, there will be two dimensionless products or π's in the solution.

To illustrate the π theorem several common equations are derived here by the methods of dimensional analysis.

Example 5-3. Let it be required to derive an expression for the distance attained by a body falling freely from rest under the influence of gravity. It is known that the distance fallen s is dependent upon the acceleration of gravity g and the time of fall t. It is assumed that a complete dimen-

---

* The π theorem is attributed to E. Buckingham. For proof of it see Bridgman's "Dimensional Analysis," Yale University Press, New Haven, Conn., 1931.

† A complete homogeneous equation is one that expresses completely the relationship among physical quantities, is independent of the size of the units used, and is dimensionally homogeneous in that every term has the same dimensions. For example, the motion of a body falling freely from rest is completely defined by the familiar equation $s = \frac{1}{2}gt^2$. This equation is dimensionally homogeneous in that both its terms have the dimension of length. Furthermore, this equation is independent of the size of the units used. In a similar manner the expression for the velocity of this body, $v = gt$, is dimensionally homogeneous and is independent of the size of units used. However, if the two equations are added, the result is

$$s + v = \frac{1}{2}gt^2 + gt$$

This equation is not dimensionally homogeneous, since length and velocity terms appear on both sides of the equation. Such equations may be perfectly correct, but they have no place in the application of the π theorem.

‡ A dimensional constant is a coefficient that does not change in value when the other physical quantities in the relationship are changed but that does change in magnitude when the measuring units are changed. The acceleration of gravity g is such a dimensional constant.

sionally homogeneous equation of the form $f(s, g, t) = 0$ may be written. The dimensional formulas of s, g, and t are L, LT^{-2}, and T, respectively. By the π theorem it is seen that since there are three quantities expressed in terms of two fundamental dimensions, then one π term may be expected. The general expression* for π is

$$\pi = s^a g^b t^c$$

and in terms of the fundamental dimensions

$$\pi = L^a(LT^{-2})^b T^c$$
or
$$\pi = L^{a+b}T^{-2b+c}$$

Since π must be dimensionless, then the exponent of each dimension in the right-hand member of this equation must be zero. Thus, the equations

$$a + b = 0$$
$$-2b + c = 0$$

must be satisfied. Two equations and three unknowns are involved. One unknown may be chosen arbitrarily if it is independent of the others. The independence of one of the unknowns is established if the determinant of the coefficients of the remaining terms does not vanish. In this problem it is desired to determine the expression for the distance s in terms of the other variables, and it is therefore logical to let $a = 1$. If the determinant of the coefficients of the b and c terms in the simultaneous equations is written, it is seen that the determinant is not zero and therefore a is independent of b and c. For $a = 1$, the solution of the simultaneous equations is $b = -1$, $c = -2$. Thus, by the π theorem,

$$\pi_1 = sg^{-1}t^{-2}$$

and since $\phi(\pi_1) = 0$ for all values of π_1, then π_1 must be a constant; therefore,

$$s = \text{const } gt^2$$

It is seen from this example that dimensional analysis has led to the form of the equation but has not yielded the value of the constant. This must be found by experiment or by some other type of analysis. In this case the constant is known to be $\frac{1}{2}$, and the complete equation is $s = \frac{1}{2}gt^2$.

In the example just used it was desired to determine an equation for s in terms of the other variables, and therefore the value of a, the exponent of s, was chosen as unity. There are, however, some restrictions on the arbitrary choice of values for the exponents. Some of these restrictions are pointed out in the following example.

* The underlying rule in dimensional analysis is that the dimensional formula of every measurable quantity is expressible as the product of the powers of the fundamental quantities upon which it depends.

Example 5-4. To determine the velocity of sound in a gas it may be assumed that the velocity v is dependent upon the pressure p, the mass per unit volume ρ_m, and the absolute viscosity μ_m. If M, L, and T are chosen for the fundamental dimensions, then the dimensional formulas of all the physical quantities involved are:

Quantity	Symbol	Dimensional formulas
Velocity...............	v	LT^{-1}
Pressure................	p	$ML^{-1}T^{-2}$
Density (mass)...........	ρ_m	ML^{-3}
Viscosity (absolute)........	μ_m	$ML^{-1}T^{-1}$

Since there are four quantities expressed in terms of three fundamental dimensions, there will be one dimensionless group or one π term in the result. This term may be written

or
$$\pi = v^a p^b \rho_m{}^c \mu_m{}^d$$
$$\pi = (LT^{-1})^a (ML^{-1}T^{-2})^b (ML^{-3})^c (ML^{-1}T^{-1})^d$$

Since π must be dimensionless, the equations

$$b + c + d = 0$$
$$a - b - 3c - d = 0$$
$$-a - 2b - d = 0$$

must be satisfied. Since an expression for the velocity v is desired, it is logical to let $a = 1$. Then $b = -\frac{1}{2}$, $c = \frac{1}{2}$, and $d = 0$. Thus,

$$\pi_1 = v p^{-\frac{1}{2}} \rho_m{}^{\frac{1}{2}}$$

and since $\phi(\pi_1) = 0$ for all values of π_1, then

$$v = \text{const} \sqrt{\frac{p}{\rho_m}}$$

The velocity, apparently, is independent of the viscosity. In this example, if it had been assumed that the viscosity would appear in the final result and the value of d had been chosen as unity, then the three simultaneous equations would have no solution, since the determinant of the coefficients of the terms without d is

$$\begin{vmatrix} 0 & 1 & 1 \\ 1 & -1 & -3 \\ -1 & -2 & 0 \end{vmatrix} = 0$$

The fact that there is no solution for the simultaneous equations if $d = 1$ indicates that the assumption that the result depends upon the viscosity is erroneous. In this way it may be possible, by the methods of dimensional analysis, to avoid the error of assuming that a result depends upon a certain physical quantity that at first thought appears to be involved but in reality is not. Superfluous terms, however, are not always identified as such.

In each of the two foregoing examples only one dimensionless group was found in the result. Frequently, however, the number of quantities involved exceeds the number of fundamental dimensions by more than one, and therefore more than one dimensionless group is expected. In such cases either it is necessary to assign arbitrary values to more than one unknown, in solving the simultaneous equations of the exponents, or it is necessary to determine some of the unknowns in terms of the others. In either case those unknowns which are assigned arbitrary values or which are used to express the values of the other unknowns must be independent of the others. An example will illustrate this type of problem.

Example 5-5. An expression for the pressure gradient (pressure loss per foot length of pipe) in a smooth, straight pipe of circular cross section is desired. The pressure gradient G is assumed to depend upon the velocity v of the fluid in the pipe, the diameter D of the pipe, the mass density ρ_m, and the absolute viscosity μ_m. It is also assumed that the expression may be written in the form $f(G, v, D, \rho_m, \mu_m) = 0$. The dimensional formulas of the various quantities in terms of F, L, and T as the fundamental dimensions are as follows:

Quantity	Units (English system)	Symbol	Dimensional formulas
Pressure gradient...............	lb$_f$/(sq ft)(ft)	G	FL^{-3}
Velocity.......................	ft/hr	v	LT^{-1}
Diameter......................	ft	D	L
Density (mass).................	slugs/cu ft	ρ_m	FT^2L^{-4}
Viscosity (absolute).............	slugs/(ft)(hr)	μ_m	FTL^{-2}

Since there are five quantities and three fundamental dimensions, then two π terms may be expected. The general equation for the π terms is

$$\pi = G^a v^b D^c \rho_m{}^e \mu_m{}^f$$

or $$\pi = (FL^{-3})^a (LT^{-1})^b L^c (FT^2L^{-4})^e (FTL^{-2})^f$$

from which the equations

$$a + e + f = 0$$
$$-3a + b + c - 4e - 2f = 0$$
$$-b + 2e + f = 0$$

are obtained. Since an expression for G in terms of the other variables is desired, it is logical to let $a = 1$. One of the other variables may be chosen equal to zero, since there are going to be two dimensionless groups in the solution. In this case let $f = 0$. If these values are substituted in the simultaneous equations, the solution is $a = 1$, $b = -2$, $c = 1$, $e = -1$, $f = 0$, and therefore

$$\pi_1 = Gv^{-2}D\rho_m{}^{-1}$$

or

$$\pi_1 = \frac{GD}{v^2\rho_m}$$

To find the second dimensionless group π_2, another set of values must be assigned to two of the unknowns in the three simultaneous equations. A value of $a = 0$ must be chosen, since π_1 already contains the term G and it would be undesirable to have G appear in more than one dimensionless group. An arbitrarily assigned value of $b = 1$ then leads to the result $a = 0$, $b = 1$, $c = 1$, $e = 1$, and $f = -1$. When these values are substituted in the equation for π, the result is

$$\pi_2 = \frac{vD\rho_m}{\mu_m}$$

From the π theorem it is seen that $\phi(\pi_1, \pi_2) = 0$, and in this case

$$\phi\left(\frac{GD}{v^2\rho_m}, \frac{vD\rho_m}{\mu_m}\right) = 0$$

This can be solved for G, since the function is zero for all values of $GD/v^2\rho_m$ and $vD\rho_m/\mu_m$. The solution is

$$\frac{GD}{v^2\rho_m} - f\left(\frac{vD\rho_m}{\mu_m}\right) = 0$$

or

$$G = \frac{v^2\rho_m}{D} f\left(\frac{vD\rho_m}{\mu_m}\right) \tag{5-9}$$

where f is some function of $vD\rho_m/\mu_m$ that must be determined by experiment.

Another method of attacking this same problem would be to write an expression for G in terms of v, D, ρ_m, and μ_m. Thus

$$G = \text{const } (v^bD^c\rho_m{}^e\mu_m{}^f)$$

or in terms of the dimensions

$$FL^{-3} = (LT^{-1})^bL^c(FT^2L^{-4})^e(FTL^{-2})^f$$

which becomes

$$FL^{-3} = F^{e+f}L^{b+c-4e-2f}T^{-b+2e+f}$$

The exponents of like dimensions on each side of the equation may be equated, since it must be dimensionally homogeneous. Hence,

$$\begin{aligned} e + f &= 1 \\ b - 4e + c - 2f &= -3 \\ -b + 2e + f &= 0 \end{aligned}$$

In this set there are three equations and four unknowns. All the unknowns may be determined in terms of f. This gives $b = 2 - f$, $c = -1 - f$, $e = 1 - f$. When these values are substituted in the original expression of G in terms of the other variables, the result is

$$G = \text{const } (v^{2-f}\rho_m{}^{1-f}D^{-1-f}\mu_m{}^f)$$

or

$$G = \text{const } \frac{v^2\rho_m}{D} \left(\frac{\mu_m}{vD\rho_m}\right)^f$$

which may be written

$$G = \text{const } \frac{v^2\rho_m}{D} \left(\frac{vD\rho_m}{\mu_m}\right)^{-f} \qquad (5\text{-}10)$$

Equation (5-10) is seen to be another form of Eq. (5-9), involving the same dimensionless groups of terms.

The examples used to illustrate the π theorem show that the methods of dimensional analysis can be used to derive equations, although the constants and the form of the functions are not given. These must be determined by direct experiment. It is often convenient to perform the experiment on a model of the apparatus rather than the full-scale apparatus. By the methods of dimensional analysis the performance of the full-scale apparatus can be predicted from the experimental knowledge gained in the model test. To predict the action of the model the principle of similarity must be applied to ensure the proper ratio of linear dimensions, forces, velocities, etc., in the model to the dimensions, forces, velocities, etc., in the full-scale apparatus.[3] The principle of similarity states that two systems are similar if the linear dimensions, forces, velocities, etc., stand in a fixed ratio to one another. To illustrate the application of dimensional analysis to model studies let it be assumed that the pressure gradient in a large-diameter pipe is to be determined by studying the pressure gradient in a small-scale model of this pipe. By the π theorem the equation for the pressure gradient is given as in Eq. (5-9). This is

$$G = \frac{v^2\rho_m}{D} f\left(\frac{vD\rho_m}{\mu_m}\right)$$

The equation for the model pipe would be

$$G_1 = \frac{v_1{}^2\rho_{1m}}{D_1} f\left(\frac{v_1D_1\rho_{1m}}{\mu_{1m}}\right)$$

The ratio of the pressure gradient in the large pipe to the pressure gradient in the small pipe is

$$\frac{G}{G_1} = \frac{(v^2\rho_m/D)f(vD\rho_m/\mu_m)}{(v_1{}^2\rho_{1m}/D_1)f(v_1D_1\rho_{1m}/\mu_{1m})}$$

The form of the function f is not known, but it must be the same in both numerator and denominator and will therefore be eliminated if

$$\frac{v_1 D_1 \rho_{1m}}{\mu_{1m}} = \frac{v D \rho_m}{\mu_m}$$

If the same fluid is used in the model that is to be used in the full-scale pipe, then $\rho_{1m} = \rho_m$ and $\mu_{1m} = \mu_m$; hence

$$v_1 D_1 = v D$$

The velocity of the fluid to be used in the model must therefore be equal to the velocity in the full-scale pipe multiplied by the ratio of the diameter of the full-scale pipe to the diameter of the model pipe; i.e., $v_1 = v(D/D_1)$. The ratio of the pressure gradients, in this case, must be

$$\frac{G}{G_1} = \frac{v^2 D_1}{v_1^2 D}$$

or since $v/v_1 = D_1/D$,

$$\frac{G}{G_1} = \left(\frac{D_1}{D}\right)^3$$

In this example the model and the full-scale pipe are similar if the velocity in the model is equal to the velocity in the full-scale pipe multiplied by the ratio of the diameters of the full-scale pipe to the model pipe, and the pressure gradient ratio is therefore inversely proportional to the cube of the diameter ratio. The following example will illustrate this type of problem.

Example 5-6. It is desired to establish the pressure gradient for air flow at a velocity of 1,000 fpm in a 16-in. pipe from tests of air flow in an 8-in. pipe. The velocity v_1 in the 8-in. pipe must therefore be vD/D_1, or $1,000 \times 1\frac{6}{8}$, equals 2,000 fpm.

A test of the 8-in. pipe at a velocity of 2,000 fpm shows the pressure gradient to be 1.0 in. of water per 100-ft length. The ratio of the pressure gradients in the 16- and 8-in. pipes, G/G_1, will be

$$\left(\frac{D_1}{D}\right)^3 = \left(\frac{8}{16}\right)^3 = \frac{1}{8}$$

The pressure gradient G at the velocity of 1,000 fpm in the 16-in. pipe equals $G_1(D_1/D)^3$, or

$$1 \times \frac{1}{8} = 0.125 \text{ in. of water per 100-ft length}$$

Problems

1. Determine the viscosity of kerosene at 80°F by use of Fig. A-6. Express the answer in lb_f-sec/sq ft.

2. Determine the viscosity of mercury at 70°F and 14.7 psia. Express the answer in lb_m/(ft)(hr).

3. It is assumed that the force F_c required to cause a particle to travel in a circular path is a function of the mass of the particle m, the radius of the path r, and the angular velocity ω. Derive, by dimensional analysis, an expression for the force F_c in terms of m, r, and ω.

4. Tests of a certain type of centrifugal pump indicated that the horse-power developed Hp depended upon the tip velocity of the impeller blades V, the impeller diameter D, and the pressure P. Determine by dimensional analysis an expression for Hp in terms of these variables.

5. Consider a wave advancing on deep water under the action of gravity; thus the velocity of the wave will depend on the acceleration of gravity, the density of the medium, and the wavelength, expressed in feet. By dimensional analysis find the arrangement of the variables in an equation for the velocity.

6. Assuming that the kinetic energy of a body is a function of its mass and its velocity, develop by dimensional analysis an expression for kinetic energy, using dimensions of F, L, and T.

7. Derive by dimensional analysis an expression for the distance S between the initial and final positions of a body when it is known that this distance is some function of the initial velocity V, the time t, and the acceleration g. It is assumed that a complete dimensionally homogeneous equation of the form $f(S, V, g, t) = 0$ can be written.

8. It is assumed that the height to which a liquid will rise in a capillary tube depends on the specific weight of the liquid, the diameter of the tube, the temperature of the liquid, and the surface tension. Surface tension has the units of pounds (force) per foot of wetted perimeter at the upper surface of the liquid in the tube. By dimensional analysis find the arrangement of the variables in an equation for the height.

REFERENCES

1. Fourier, J. B.: "Théorie analytique de la chaleur," p. 391, Gauthier-Villars, Paris, 1822; English translation by Freeman, Cambridge, 1878.
2. Bridgman, P. W.: "Dimensional Analysis," Yale University Press, New Haven, Conn., 1931.
3. Langhaar, H. L.: "Dimensional Analysis and Theory of Models," pp. 60–75, John Wiley & Sons, Inc., New York, 1951.

CHAPTER 6

FLUID FLOW IN THE CONVECTION PROCESS

6-1. General. Fluid flow, in its various aspects, is a broad subject. The fundamental laws and principles of the subject apply wherever fluids are utilized, as in chemical and industrial processes, for water and gas supply, disposal of waste, transportation, power generation, heating, cooling, and for other purposes. Other phases of the subject apply more specifically to the individual fields in which fluids are utilized. The following study of fluid flow will be limited to a review of some of the fundamental laws and principles and to certain phases which have a direct bearing upon the process of heat transfer by convection. Fluid flow, in general, deals with both compressible and incompressible fluids. Consideration will be given here only to incompressible fluids, i.e., to liquids, and to compressible fluids under conditions where the compressibility effect is so slight that they may properly be treated as incompressible fluids—conditions which are generally considered to be met by gases subjected to pressure changes of not more than about 7 per cent of the initial absolute pressure.

6-2. Flow of an Ideal Fluid. The general energy equation for one-dimensional steady-flow processes, with which all students of elementary thermodynamics should be familiar, establishes the energy stored in a given mass of fluid at any point along its path as the sum of:

1. The potential energy due to elevation.

2. The kinetic energy, which depends upon velocity.

3. The internal energy, which depends upon the temperature and state of the fluid.

4. Flow work, which is the energy required to force the given volume of fluid into or out of the system at the existing pressure.

Increase or decrease in the energy occurs if:

5. Heat is added to or removed from the fluid or if

6. Work is done on or by the fluid.

In a study of fluid flow a specialized form of the general energy equation, known as Bernoulli's equation, is of value in pointing out some fundamental relationships. This equation is generally, although not necessarily, limited to consideration of items 1, 2, and 4 of the general energy equation as listed above, in that it is usually based upon the assumption of constant temperature and state of the fluid, and therefore no change in internal energy, and upon the assumptions of no heat transfer and no performance of work. The assumption that no work is done on or by the fluid means that there is no loss of pressure due to friction; i.e., the effect of viscosity is neglected. In this sense the fluid is considered ideal and not real. The assumption of no loss of pressure due to friction, however, in no way limits the change in pressure due to the difference in velocity which occurs when the cross-sectional area of the stream is changed. The equation in this restricted form is

$$\frac{gz_1}{g_c} + \frac{V_1{}^2}{2g_c} + p_1v_1 = \frac{gz_2}{g_c} + \frac{V_2{}^2}{2g_c} + p_2v_2 \qquad (6\text{-}1)$$

where the subscripts 1 and 2 denote two different positions along the flow path. Each of the terms represents energy per unit mass. If the unit of mass is the pound, then all terms of the equation are in foot-pounds per pound mass, ft-lb$_f$/lb$_m$.

$\dfrac{gz}{g_c}$ = potential energy due to elevation, where z is the elevation in feet above any chosen datum level.

$\dfrac{V^2}{2g_c}$ = kinetic energy. If V^* is the velocity, ft/sec, then g_c is 32.1739 (lb$_m$/lb$_f$) ft/sec^2.

pv = flow work—the product of the pressure p, lb/sq ft, and the specific volume v, cu ft/lb$_m$. If the equation is limited in its application to isothermal incompressible fluids, then the specific volumes v_1 and v_2 are equal.

In experimental studies of fluid flow the energy at any position along the flow path is not measured directly; instead, measure-

---

* The velocity of an ideal fluid would be uniform throughout the cross-sectional area of the stream at any position along its path, in keeping with the assumption that no friction is present.

ments are made of the pressures, which are related to the various forms of energy—potential energy, kinetic energy, and flow work. An expression corresponding to Eq. (6-1), applicable to incompressible fluids, but in terms of pressure instead of energy, is therefore desirable and is obtained by multiplying each term by the density of the fluid ρ, lb_m/cu ft. The resulting equation is

$$\frac{g\rho z_1}{g_c} + \frac{\rho V_1^2}{2g_c} + p_1 = \frac{g\rho z_2}{g_c} + \frac{\rho V_2^2}{2g_c} + p_2 \qquad (6\text{-}2)$$

where $\dfrac{g\rho z}{g_c}$ = static pressure due to elevation above the chosen datum level, lb/sq ft.

$\dfrac{\rho V^2}{2g_c}$ = velocity pressure, lb/sq ft. If the velocity V is expressed in feet per second, then $g_c = 32.1739$ (lb_m/lb_f) ft/sec².

p = static pressure required to force each pound mass of the fluid into or out of the system, lb/sq ft.

The static pressure, which at any point along the flow path is exerted equally in all directions, is the sum of $g\rho z/g_c$ and p. In dealing with the flow of gases, however, the changes in the value of the first of these two terms are ordinarily so small as to be of no significance and the term p, alone, represents the static pressure. The velocity pressure $\rho V^2/2g_c$, since it is dependent directly upon the velocity, is exerted in the direction of flow. The *total pressure*, or *stagnation pressure*, at any position is the sum of the static pressure and the velocity pressure; or, more specifically, it is the *algebraic* sum of the static pressure and the velocity pressure, for, at times, the static pressure when measured as a gauge pressure may be negative, i.e., below the pressure of the atmosphere as a datum. Measurement of the velocity pressure as the difference between the total pressure and the static pressure serves as a means of determining the velocity.

Figure 6-1 shows the changes in static pressure and velocity pressure which occur with changes in elevation and cross-sectional area of the conduit through which an ideal (frictionless) fluid flows. At positions of different cross-sectional area along the conduit the velocities, and consequently the velocity pressures, must differ in order that a fixed quantity of fluid may continuously pass each position. The velocities must satisfy the law of continuity, which in equation form is stated variously as:

$$\text{Volume flow rate } Q = AV \qquad \text{cu ft/unit of time} \qquad (6\text{-}3)$$
$$\text{Weight flow rate } w = \gamma AV \qquad \text{lb}_f/\text{unit of time} \qquad (6\text{-}4)$$

And on the basis of flow per unit area:

$$\text{Mass flow rate } G = \rho V \qquad \text{lb}_m/(\text{sq ft})(\text{unit of time}) \ (6\text{-}5)$$

The pressure which a column of fluid of a given height exerts in all directions at its base is usually expressed in pounds force per unit area; however, it may be described otherwise by the height of the column of the fluid itself, in which case it is termed

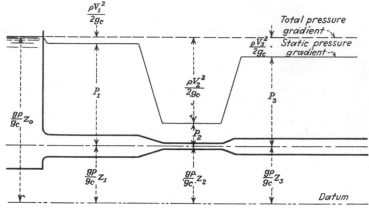

FIG. 6-1. Pressure changes for flow of an ideal (frictionless) incompressible fluid through a horizontal tube with variation in cross-sectional area.

the head and is expressed commonly in feet of the fluid. Accordingly, if desired, Eq. (6-2) may be expressed in terms of head instead of pressure by multiplying each term by $g_c/g\rho$ or by its equivalent, $1/\gamma$ [shown by Eq. (5-6)]. The resulting equation is

$$z_1 + \frac{V_1{}^2}{2g} + \frac{p_1}{\gamma} = z_2 + \frac{V_2{}^2}{2g} + \frac{p_2}{\gamma} \qquad (6\text{-}6)$$

where z = static head due to elevation above the chosen datum level, ft of fluid.

$\dfrac{V^2}{2g}$ = velocity head, ft of fluid.

$\dfrac{p}{\gamma}$ = static head required to force each pound having a specific weight γ into or out of the system, ft of fluid.

Example 6-1.   Referring to Fig. 6-1, applied to the flow of water at 70°F, assume z_0 to be 50 ft, and g and g_c both to be numerically equal to 32.17. If the inside diameters of the tube at positions 1, 2, and 3 are 6, 2, and 4 in., respectively, determine the velocity pressure and the static pressure at each of the three positions for a flow of 360 gpm.

Solution

$$\text{Flow of water, cu ft/sec} = \frac{360}{60} \times \frac{231}{1,728} = 0.802 \text{ cfs}$$

The cross-sectional areas at positions 1, 2, and 3 are 0.1965, 0.0218, and 0.0872 sq ft, respectively.   By application of the continuity equation, $Q = AV$, the velocities Q/A at positions 1, 2, and 3 are 4.08, 36.79, and 9.20 fps, respectively.   The velocity pressure $\rho V^2/2g_c$, at position 1, equals $(62.27 \times 4.08^2)/(2 \times 32.17) = 16.1$ lb/sq ft, or 0.11 lb/sq in.

By similar calculations

Velocity pressure at position 2 = 1,310.0 lb/sq ft, or 9.1 lb/sq in.
Velocity pressure at position 3 = 82.0 lb/sq ft, or 0.57 lb/sq in.

The total pressure above the datum shown is

$$\frac{g\rho z_0}{g_c} = 62.27 \times 50 = 3,113.5 \text{ lb/sq ft}$$

The static pressure at each position is the total pressure minus the velocity pressure.

Static pressure at position 1 = 3,113.5 − 16.1 = 3,097.4 lb/sq ft,
or 21.5 lb/sq in.
Static pressure at position 2 = 3,113.5 − 1,310.0 = 1,803.5 lb/sq ft,
or 12.5 lb/sq in.
Static pressure at position 3 = 3,113.5 − 82.0 = 3,031.5 lb/sq ft,
or 21.0 lb/sq in.

6-3. Flow of a Real Fluid.   A real fluid differs from the ideal, which has up to this point been considered, in that it possesses viscosity, a property which may be briefly described as a resistance to flow.   Interaction of adjacent molecules within the fluid (cohesion) and between molecules of the fluid and of the solid which confines it (adhesion) combines to resist motion. When one layer of fluid is forced to move past an adjacent layer of fluid or of solid, the interaction of molecules at the boundary tends to check the relative motion.   All fluids, whether liquid or gaseous, offer resistance to the shearing stress which is created between adjacent layers, i.e., possess the property called viscosity. In general, for liquids, the viscosity decreases with rise in temperature; for gases it increases.

When the viscosity of the fluid is taken into consideration, then the frictional resistance to flow must be accounted for.   In this case, a certain portion of the total energy and corresponding total pressure of the fluid at any position in the flow path is used in overcoming the frictional resistance which is encountered between that position and any position farther downstream. Equation (6-2) accordingly may be modified by the inclusion of a term to represent the loss of total pressure due to frictional resistance.   The equation then becomes

$$\frac{g\rho z_1}{g_c} + \frac{\rho V_1{}^2}{2g_c} + p_1 = \frac{g\rho z_2}{g_c} + \frac{\rho V_2{}^2}{2g_c} + p_2 + \Delta p \qquad (6\text{-}7)$$

where Δp = loss of total pressure, lb/sq ft, and all other terms are as in Eq. (6-2).

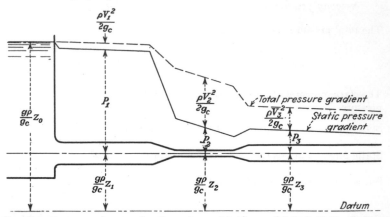

FIG. 6-2. Pressure changes for flow of a real incompressible fluid (possessing viscosity) through a horizontal tube with variation in cross-sectional area.

Figure 6-2 shows the changes in total pressure, velocity pressure, and static pressure which occur when a real fluid flows through a conduit which varies in cross-sectional area at different positions along its path.   It may be noted that in the case of flow through any portion of the system where the cross-sectional area remains uniform and where the velocity pressure necessarily remains constant, the loss of total pressure is also equal to loss of static pressure.

At all velocities, resistance to flow is encountered in varying degree in paths of different cross section and shape, primarily because of differences which occur in the velocities of particles or layers of fluid in contact with each other. Whenever the flow is guided or directed, as when a liquid or gas flows within a pipe, a layer of fluid adheres to the pipe wall and remains practically at rest, even though the fluid at the central axis of the pipe moves at a high velocity. The pattern of the flow, as characterized by the variation in velocity across the pipe, however, is dependent upon the average velocity of the stream. At very low velocities and at a constant temperature each particle of the fluid moves in a path substantially parallel to the path followed by all other particles and the flow is described as *streamline*, or *laminar*. This does not mean that each particle moves at the same velocity as every other particle, but simply that all particles travel in parallel directional paths. In fact, the average

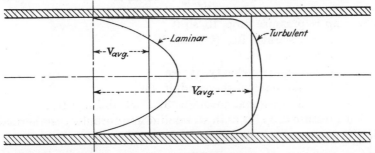

Fig. 6-3. Comparison of the velocity-distribution curves in laminar and turbulent flow.

velocity over the entire cross section for isothermal flow is only 0.5 of the maximum that occurs at the center of the stream. A plot of the variation in velocity across the pipe, as shown in Fig. 6-3, takes the shape of a parabola. If the temperature of the fluid across the pipe changes because of heat transfer to or from the pipe wall, the velocity gradient across the pipe becomes somewhat distorted from the parabolic shape and the flow is characterized as nonisothermal, or modified laminar, flow.

When the flow is increased beyond a certain critical velocity, streamline flow can no longer continue and turbulence takes place. In this range of *turbulent* flow, innumerable eddies and cross currents occur in the main body of the stream. The film which adheres to the surface is now in the form of two layers: the first, or sublayer, composed of particles completely without motion clinging to the surface and particles creeping along in streamline flow with increasing velocity as the distance from the surface is increased; and the second layer, much thicker than the first, being a transition zone composed of eddy currents moving at a higher velocity although not so swiftly as the main portion of the fluid stream. The boundary between these two layers is not sharply defined. The average velocity over the entire cross section is now of the order of 0.8 or 0.9 of the maximum velocity, the ratio in general increasing with the equivalent diameter of the conduit and the smoothness of the inner surface.

In streamline flow certain mathematical relations hold among the pressure drop, the rate of flow, the viscosity of the fluid, and other factors, as expressed by the Poiseuille formula

$$\Delta p = \frac{32\mu L V}{g_c D^2} \quad \text{lb/sq ft} \tag{6-8}$$

where Δp = pressure drop, lb/sq ft.

μ = viscosity, $lb_m/(ft)(sec)$.*

L = length, ft.

V = average velocity, ft/sec.

D = diameter of the conduit, ft.

g_c = dimensional constant, 32.1739 (lb_m/lb_f) ft/sec².

The pressure drop for both streamline and turbulent isothermal flow is expressed by the Darcy-Weisbach formula

$$\Delta p = \frac{fL\rho V^2}{8g_c m} \quad \text{lb/sq ft} \tag{6-9}$$

where f = friction factor.

ρ = density of the fluid, lb_m/cu ft.

m = hydraulic radius, i.e., the cross section of the conduit divided by the wetted perimeter ($D/4$ for circular sections).

All other terms are the same as in Eq. (6-8).

The application of Eqs. (6-8) and (6-9) for the purpose of determining the pressure drop encountered in fluid flow will be discussed later in this chapter. Here, these two equations are made to serve only as a means for establishing the critical velocity. Knowledge of the critical velocity is necessary, since the laws of heat transfer applying to streamline and to turbulent flow are not the same.

* Any unit of time may be employed provided the same unit is applied to V.

The transition that occurs at the critical velocity from streamline to turbulent flow is neither instantaneous nor positively fixed. Any disturbance of the fluid in the neighborhood of the critical velocity, such as may be produced by jarring the conducting pipe, may change streamline into turbulent flow. As a result there is an appreciable overlapping of the highest velocity for streamline flow and the lowest velocity at which turbulent flow may occur. In the range of streamline flow the frictional resistance is practically independent of the roughness of the surface, and in the range of usual pipe sizes it is only slightly dependent upon the pipe diameter; but in turbulent flow both these factors influence the pressure drop. However, if the two equations (6-8) and (6-9) which apply at both boundaries of the critical-velocity zone for flow in a pipe are equated, an approximate value of the critical velocity may be found, expressed as

$$V_{crit} = \frac{64\mu}{f\rho D} \quad \text{ft/sec} \tag{6-10}$$

The value of f for the highest velocity in the range of streamline flow has been established as very close to 0.018. For the lowest velocity in turbulent flow, f for smooth pipe equals 0.044; it is slightly higher if the pipe surface is rough. Substitution of an average value of 0.031 for f in Eq. (6-10) shows that the critical velocity for flow in a pipe may be expressed approximately by the equation

$$V_{crit} = \frac{2,100\mu}{\rho D} \quad \text{ft/sec} \tag{6-11}$$

Since the values of μ and ρ vary with temperature and are both different for different substances, it is obvious that even for flow in a rigid pipe of fixed diameter the critical velocity V is not always the same; but an average value of the product $DV\rho/\mu$ may be considered constant and of the order of 2,100. The fraction $DV\rho/\mu$ has been given the name *Reynolds number* because of its early mention by Osborne Reynolds[1] in 1883 and is often written N_{Re}. The units of the terms in the numerator, expressed on any time basis such as the hour, are

$$\text{ft} \times \text{ft/hr} \times \text{lb}_m/\text{cu ft} = \text{lb}_m/(\text{ft})(\text{hr})$$

which are the same units in which the denominator μ is expressed. The number is accordingly dimensionless. In hydraulic studies great significance is attached to the Reynolds number, since the frictional resistance encountered by various fluids flowing at constant temperature in any pipe having a known condition of surface roughness has been found to depend upon the Reynolds number only.

The Reynolds number of approximately 2,100 at the critical velocity in pipes has been verified substantially by a number of experimental methods, such as the observance of the motion of colloidal particles in a liquid flowing in a glass pipe, and by the use of a stethoscope to detect the change in noise at the velocity where turbulent flow begins. These experiments indicate that the Reynolds number at the critical velocity may range from approximately 2,100 to 3,100; so for a Reynolds number below 2,100 the flow may be considered streamline, and for a number above 3,100 the flow is considered turbulent.

An idea of the magnitude of critical velocities may be gained by application of Eq. (6-11) to a few specific examples such as the calculation of the critical velocity of water, air at atmospheric pressure, and light motor oil, each at 70°F, flowing in a 2-in. pipe (actual diameter 2.067 in., or 0.172 ft).

For water, the critical velocity is

$$V_{crit} = \frac{2,100\mu}{\rho D} = \frac{2,100 \times 2.37}{62.3 \times 0.172}$$
$$= 464 \text{ ft/hr, or } 0.129 \text{ ft/sec}$$

For air,

$$V_{crit} = \frac{2,100 \times 0.044}{0.075 \times 0.172}$$
$$= 7,180 \text{ ft/hr, or } 2.0 \text{ ft/sec}$$

For oil,

$$V_{crit} = \frac{2,100 \times 242}{54 \times 0.172}$$
$$= 54,600 \text{ ft/hr, or } 15.2 \text{ ft/sec}$$

For flow in larger pipes, or at higher fluid densities, the critical velocity is less, in inverse proportion to the pipe diameter and the density of the fluid.

A comparison of the critical velocities found in the foregoing examples suggests that the majority of problems in heat transfer to or from water and air in pipes are in the range of turbulent flow whereas many problems dealing with oils are in the range of streamline flow.

6-4. Pressure Loss in Pipes and Tubes. Equations (6-8) and (6-9) were presented in Art. 6-3, showing the factors which influence the pressure loss in pipes and tubes for streamline and turbulent flow. No attempt was made at that point to apply these equations for determining the magnitude of pressure losses. A knowledge of the magnitude of pressure losses is important, however, in the design of many forms of heat exchangers, not only because of the influence of pressure loss upon the power required to circulate a fluid, but also because, as will be shown later, a relationship exists between the pressure loss and the transfer of heat between the tube wall and the fluid stream.

In applying Eqs. (6-8) and (6-9) to the calculation of pressure loss it should be observed that Eq. (6-8) applies only to streamline flow, whereas Eq. (6-9) applies to either streamline or turbulent flow. For flow in a pipe or tube of circular cross section, Eq. (6-9) may be more conveniently expressed as

$$\Delta p = f \frac{L\rho V^2}{D2g_c} \quad \text{lb/sq ft} \tag{6-12}$$

or the loss in head

$$h_f = f \frac{LV^2}{D2g} \quad \text{ft of fluid} \tag{6-13}$$

Calculations of the loss in pressure or head may be made by substituting into one or the other of these expressions appropriate values of the friction factor f, the length L, and the density ρ, in addition to values of the tube diameter and the velocity. If the velocity is expressed in feet per second, the value of g is 32.1739 ft/sec^2; but if the velocity is expressed in feet per hour, the value of g is 4.17×10^8 ft/hr^2.

Equation (6-13), when transposed, shows that the friction factor $f = h_f/(L/D)(V^2/2g)$, but since $h_f/(L/D)$ represents the loss in head for a length of tube equal to one diameter, it is apparent that the friction factor f may be found experimentally as the ratio of the loss in head per length of one diameter to the velocity head $V^2/2g$.

Values of the friction factor f are shown in Fig. 6-4 corresponding to calculated values of the Reynolds number. This friction factor applies to the Darcy-Weisbach equation in the forms shown

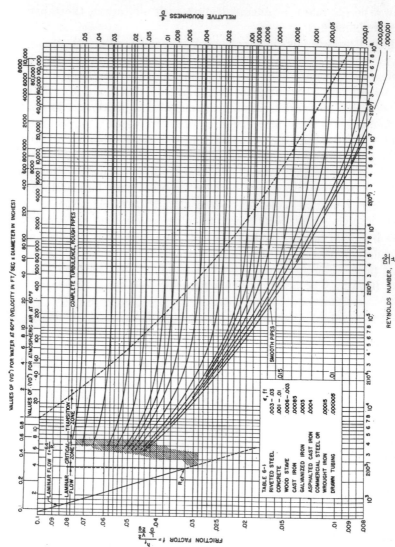

FIG. 6-4. Relation between friction factor and Reynolds number. [L. F. Moody, Friction Factors for Pipe Flows, Trans. ASME, 66, 671–694 (1944).]

in Eqs. (6-9), (6-12), and (6-13) and is four times as large as the corresponding friction factor f_F applied to the Fanning equation, $\Delta p = 4f_F(L\rho V^2/D2g_c)$, which appears in many heat-transfer publications. If the flow is isothermal, or substantially so, the Reynolds number is based upon evaluations of the viscosity and density at the existing temperature. When heat is transferred at such a rate that there is an appreciable difference in the temperature of the tube wall and the main body of fluid, the evaluation of viscosity should be made at a temperature which represents a weighted average. For streamline flow of viscous fluids, this temperature t' is usually taken to be equal to $(3t + t_w)/4$, where t is the temperature of the main body of fluid (often termed the bulk temperature) and t_w is the temperature of the tube wall; and for turbulent flow of viscous fluids, t' is usually considered to be the arithmetic mean of the main body and the surface temperatures. For turbulent flow of viscous fluids such as petroleum oils, for which the viscosity changes markedly with change in temperature, Sieder and Tate[3] recommend that the friction factor based upon the main-body temperature be divided by $(\mu/\mu_s)^{0.14}$, where μ and μ_s are, respectively, the viscosities at the main body and at the surface temperatures.

In Fig. 6-4, values of the friction factor f are shown not only as a function of the Reynolds number but also as a function of the relative roughness of the surface ϵ/D, ϵ being a linear dimension of the roughening protruberances, in the same units as the diameter D. Table 6-1, inserted in Fig. 6-4, shows values of ϵ, in feet, for different kinds of pipe. The shapes of the curves for different values of the Reynolds number and of relative roughness are significant. In the range of laminar flow the friction factor is seen to be dependent upon the Reynolds number alone; i.e., it is independent of the roughness of the surface. In this range it is numerically equal to $64/N_{Re}$, where N_{Re} denotes the Reynolds number. For turbulent flow through fairly smooth pipes the friction factor is dependent mainly upon the Reynolds number and only slightly upon the roughness of the surface, and the slope of the curve is great, especially at low Reynolds numbers. For flow through very rough pipes the friction factor depends primarily upon the relative roughness and is influenced to a lesser degree by the Reynolds number, thereby causing the curves to be comparatively flat. Decrease in the friction factor with increase in the Reynolds number and with decrease in the relative roughness is to be expected in keeping with the influences of these factors upon the velocity gradient across pipes of different roughness. In Fig. 6-3, typical velocity gradients are shown for both laminar and turbulent flow. The gradient for turbulent flow is seen to be much flatter than for laminar flow. If a gradient for turbulent flow were shown for a greater value of Reynolds number it would be a still flatter curve. If a gradient were shown for the same Reynolds number but for a smoother pipe it would likewise take a flatter shape, indicating in each case a higher ratio of average velocity to the maximum (or center) velocity. The flatter gradients, resulting from an increase in Reynolds number or from a decrease in roughness, indicate smaller differences in the velocities of adjacent particles and layers of fluid and therefore reduced frictional resistance.

6-5. Pressure Loss in Noncircular Conduits. The friction factor for turbulent flow in pipes has been shown to be dependent upon the Reynolds number and the relative roughness. Both these terms involve the diameter D. When either term is applied to flow through conduits of noncircular cross section, such as square and rectangular ducts and annular spaces between concentric pipes and tubes, a knowledge of the equivalent diameter is useful.

No simple satisfactory expression for the equivalent diameter has been developed for laminar flow in noncircular conduits, since this type of flow does not depend upon the relative roughness of the surface; but for turbulent flow it has been assumed and has been verified experimentally[4] that at any given average velocity, conduits having surfaces of equal roughness but of different shapes will have practically the same pressure loss in a length equal to the ratio of the cross-sectional area to the perimeter. This ratio of area to perimeter, A/P, is called the hydraulic radius; and $4A/P$ is the hydraulic or equivalent diameter,* usually denoted by D_e. For flow, either laminar or turbulent, in a pipe or tube of circular cross section,

$$D_e = \frac{4\pi D^2}{4\pi D} = D \tag{6-14}$$

---

* It has already been noted that the Darcy-Weisbach friction factor f is four times the corresponding Fanning friction factor f_F. The former factor

For turbulent flow in an annular space,

$$D_e = \frac{4\pi(D_2{}^2 - D_1{}^2)}{4\pi(D_2 + D_1)} = \frac{(D_2 - D_1)(D_2 + D_1)}{D_2 + D_1} = D_2 - D_1 \quad (6\text{-}15)$$

where D_1 = outside diameter of the inner pipe and D_2 = inside diameter of the outer pipe. For turbulent flow in a duct of square cross section of width and depth w,

$$D_e = \frac{4w^2}{4w} = w \quad (6\text{-}16)$$

For turbulent flow in a conduit of rectangular cross section with sides a and b,

$$D_e = \frac{4ab}{2(a + b)} = \frac{2ab}{a + b} \quad (6\text{-}17)$$

Huebscher[4] has found that for most practical purposes the equivalent diameter, as found by Eq. (6-17), can be applied satisfactorily to calculations of pressure loss in rectangular conduits of aspect ratio (ratio a/b) not exceeding 8:1.

It should be noted that even though the pressure loss is the same for flow at any given velocity through conduits of different shape but of the same equivalent diameter, the flow rates, which depend upon the cross-sectional area, may be quite different. The flow rate through a square duct is 27.3 per cent greater than through a round duct of the same equivalent diameter when the velocities in the two ducts are the same.

In calculations of pressure loss based upon the flow rate, Reynolds number may conveniently be expressed as

$$N_{\text{Re}} = \frac{D_e V \rho}{\mu} = \frac{4A}{P} \frac{m}{\rho A} \frac{\rho}{\mu} = \frac{4m}{P\mu} \quad (6\text{-}18)$$

where m = mass flow rate, lb_m/hr.

P = perimeter of the pipe or duct, ft.

μ = viscosity of the fluid, $\text{lb}_m/(\text{ft})(\text{hr})$.

In the form of Eq. (6-18), the Reynolds number is seen to be a function of the quantity and viscosity of the fluid and of the perimeter of the fluid stream, regardless of the shape of its cross section.

---

is based upon the loss of pressure in a length equal to the hydraulic diameter, whereas the Fanning friction factor is based upon the loss of pressure in a length equal to the hydraulic radius.

Example 6-2.   Water at 70°F flows at a velocity of 4 fps through a 1½-in. wrought-iron pipe (actual i.d. = 1.61 in.).   Determine the pressure loss in a 100-ft length of pipe, expressing the result in (*a*) pounds per square inch and (*b*) feet of water.

Solution.   (For properties of water, see Table A-1.)

$$\text{Reynolds number } \frac{DV\rho}{\mu} = \frac{1.61 \times 4 \times 3{,}600 \times 62.27}{12 \times 2.37} = 50{,}700$$

$$\text{Relative roughness } \frac{\epsilon}{D} = \frac{0.00015 \times 12}{1.61} = 0.00112$$

$$\text{Friction factor } f \text{ (Fig. 6-4)} = 0.0245$$

(*a*) By Eq. (6-12)

$$\Delta p = f\frac{L\rho V^2}{D2g_c} = \frac{0.0245 \times 100 \times 62.27 \times 4^2 \times 12}{1.61 \times 2 \times 32.17} = 282 \text{ lb/sq ft}$$

or
$$282\!\!\;/\!\!\;_{144} = 1.96 \text{ lb/sq in.}$$

(*b*) By Eq. (6-13)

$$h_f = f\frac{LV^2}{D2g} = \frac{\Delta p}{\rho}\frac{g_c}{g} = \frac{282}{62.27} = 4.52 \text{ ft of water}$$

Example 6-3.   Air at atmospheric pressure and at an average temperature of 300°F flows at the rate of 6,000 cfm through a 24- by 18-in. galvanized iron duct.   Determine the loss in static head in a 50-ft length, expressed in inches of water at 70°F.

Solution.   (For properties of air, see Table A-2.)

By Eq. (6-18)

$$N_{Re} = \frac{4m}{P\mu} = \frac{4 \times 6{,}000 \times 60 \times 0.0522}{7 \times 0.058} = 185{,}000$$

$$D_e = \frac{4A}{P} \quad \text{or} \quad \frac{2ab}{a+b} = \frac{2 \times 2 \times 1.5}{2 + 1.5} = \frac{6}{3.5} = 1.715 \text{ ft}$$

$$\frac{\epsilon}{D} = \frac{0.0005}{1.715} = 0.000292$$

$$\text{Friction factor } f \text{ (Fig. 6-4)} = 0.018$$

$$\text{Velocity, fps} = \frac{Q}{A \times 60} = \frac{6{,}000}{3 \times 60} = 33.3 \text{ fps}$$

By Eq. (6-13)

$$h_f = f\frac{LV^2}{D2g} = 0.018\,\frac{50 \times (33.3)^2}{1.715 \times 2 \times 32.17} = 9.04 \text{ ft of air}$$

$$\text{Loss in static head, in. of water} = h_f\,\frac{\rho \text{ for air} \times 12}{\rho \text{ for water}}$$

$$= 9.04 \times \frac{0.0522 \times 12}{62.27} = 0.063 \text{ in. of water}$$

6-6. Pressure Loss at Inlet Ends of Pipes and Tubes. The velocity gradients which have been described for flow in pipes and tubes apply not at the inlet of the pipe or tube but farther downstream where the flow pattern is fully developed. An understanding of the changes that take place between the inlet and the point of fully developed flow can be gained by first considering the flow pattern on either side of a thin plate held in a stationary position in a moving stream of fluid and with its surfaces parallel to the direction of flow. Because of viscosity a velocity gradient is set up within the fluid, at right angles to the direction of flow. The fluid in contact with the surface of the plate is brought to rest, and a shear stress is produced within the fluid. Although the stress is imparted to the entire stream, it is mainly concentrated in a thin layer close to the surface—a

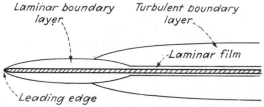

Fig. 6-5. Flow pattern along a thin plate.

layer in which there are great differences in the velocities of adjacent particles. This thin layer is known as the boundary layer. Its existence and behavior were first conceived and investigated by the German scientist Ludwig Prandtl near the beginning of the present century. Studies of the stresses and changes in energy and momentum within this layer have been the basis for the development of the boundary-layer theory[5] by which it has been possible to solve many very important and complex problems in the fields of aerodynamics, fluid mechanics, turbine design, and heat transfer.

Investigation of the flow past the thin plate in the moving stream has shown that at the leading edge of the plate the thickness of the boundary layer is zero, the thickness increasing as the distance from the leading edge increases. The flow pattern on both sides of the plate is shown in Fig. 6-5, but not to scale; here the vertical scale is much greater than the horizontal, thus magnifying the thickness of the boundary layer and the laminar film. In the thin portion of the boundary layer the flow is laminar, the velocity varying from zero at the surface of the plate to the mainstream velocity at the opposite side of the layer. If the stream moves at sufficient velocity a certain critical thickness of the boundary layer will be reached where the flow within it changes from laminar to turbulent, except within a very thin sublayer where the flow remains laminar and within the intervening transition zone (buffer layer) where the flow is neither laminar nor completely turbulent. The thickness of the boundary layer increases at a greater rate where the flow changes from laminar to turbulent, but even in the turbulent region the layer is actually very thin.

In the case of flow at the entrance to a pipe, a similar formation of a boundary layer occurs, as shown in Fig. 6-6. The boundary

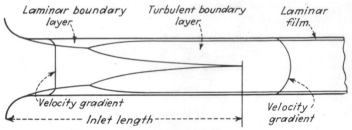

FIG. 6-6. Flow pattern at pipe entrance.

layer is first in laminar form and then, farther downstream, in turbulent form, provided of course the Reynolds number for the stream is high enough for turbulent flow. In this case, however, the turbulent boundary layer continues to increase in thickness until it eventually extends to the central axis of the pipe. Even in laminar flow, unless the pipe is unusually large, the boundary layer will eventually extend to the central axis of the pipe. When this occurs the flow pattern is fully developed. The distance from the inlet end of the pipe to the point of fully developed flow (called the inlet length) depends upon the shape of the entrance, the roughness of the pipe surface, and the Reynolds number; but an inlet length of at least 30 diameters is usually considered necessary for producing fully developed flow. In the case of laminar flow the inlet length is much more than 30 diameters. Throughout the inlet length the boundary layer is thinner than farther downstream and the differences in the velocities of adjacent particles of fluid are therefore greater.

The foregoing explanation shows that the disturbance to the normal flow pattern which is caused by the entrance to a pipe actually results in increased friction throughout the entire inlet length. The pressure loss, which is usually termed the inlet loss, is in reality the increase in pressure loss beyond that which would occur with fully developed flow. In calculations it appears as a separate item, independent of the friction loss calculated for the entire length of pipe under fully developed flow.

If the entrance to the pipe is rounded (bell-shaped), as in Fig. 6-6, the inlet loss is small, of the order of 4 per cent of the velocity pressure corresponding to the average velocity in the pipe. If, however, the inlet end of the pipe is cut square or is flush with the inner surface of any larger enclosure to which it is connected, then some separation of the boundary layer from the pipe wall takes place immediately downstream of the entrance. The stream at first contracts and then gradually expands, the smallest section (the vena contracta) occurring at about one-half of the pipe diameter downstream of the inlet. If the cross-sectional area of the pipe is small in comparison with the cross-sectional area upstream of the inlet, the area at the vena contracta may be as small as 60 per cent of the area of the pipe. The space outside the jet is filled with fluid that is practically stationary. Little loss occurs between the inlet and the vena contracta, but from there on, until the flow pattern is fully developed, considerable friction occurs. The loss has been found to be practically independent of the surface roughness but proportional to the velocity pressure and may be as high as 50 per cent of the velocity pressure. In equation form the inlet loss, which is a loss in both total and static pressure, may be expressed as

$$\Delta p = \frac{K \rho V_1{}^2}{2 g_c} \qquad \text{lb/sq ft} \qquad (6\text{-}19)$$

or the loss in head

$$h_f = \frac{K V_1{}^2}{2g} \qquad \text{ft of fluid} \qquad (6\text{-}20)$$

where V_1 = the average velocity in the pipe, ft/sec or ft/hr, according to the unit of time applied to g_c and g; and K = function of the ratio of the cross-sectional area of the pipe A_1 to that of the larger enclosure A_2 to which it is connected. Values of K are shown in Table 6-2.

TABLE 6-2. FRICTION COEFFICIENT FOR ABRUPT ENTRANCES
(From Hughes and Safford[6])

Ratio A_1/A_2	0.1	0.2	0.3	0.4	0.5	0.6	0.7	0.8	0.9	1.0
K	0.44	0.35	0.29	0.24	0.19	0.15	0.10	0.07	0.04	0.00

6-7. Pressure Loss at Enlargement in Pipes and Tubes.
Where a sudden enlargement occurs, the boundary layer of the fluid separates from the pipe surface and a jet of cylindrical shape is formed. The space outside the jet is filled with fluid that is practically stationary. Gradual enlargement of the jet starts a very short distance downstream of the point of change in cross section of the pipe and continues until the jet contacts the pipe wall. In the distance through which expansion of the jet takes place the frictional resistance is comparatively large. Analysis, by application of the principle of conservation of momentum, has shown that the pressure loss is a function not of the difference in upstream and downstream velocity pressures $\rho V_1^2/2g_c$ and $\rho V_2^2/2g_c$ but of the square of the difference in velocities. Experimental determinations show that the pressure loss caused by a sudden enlargement may be expressed with fair accuracy as

$$\Delta p = \frac{\rho(V_1 - V_2)^2}{2g_c} \qquad \text{lb/sq ft} \qquad (6\text{-}21)$$

or the loss of head

$$h_f = \frac{(V_1 - V_2)^2}{2g} \qquad \text{ft of fluid} \qquad (6\text{-}22)$$

Where an enlargement in a pipe is provided, much of the pressure loss shown by Eq. (6-21) may be avoided if the increase in cross section can be made gradual enough to prevent separation of the boundary layer from the pipe wall. In order to accomplish this the angle of divergence of the pipe wall from the axis of the pipe should be not more than about 7 deg. For a 3-deg angle of divergence the loss is only about 13 per cent of the loss for a sudden enlargement, whereas for angles of more than 7 deg the loss is substantially the same as for a sudden enlargement.

An enlargement usually occurs in a sheet-metal duct where the duct divides into two or more branches; here the sum of the cross-sectional areas of the branches commonly exceeds that of the main duct. In such a case the pressure loss is uncertain but is often estimated to be 50 per cent of the regain in static pressure produced by the reduction in velocity pressure.

It should be observed that the pressure losses for enlargements and for entrances to pipes and tubes occur independently of the conversion that takes place between static and velocity pressures, even though the pressure loss for entrances has been expressed in Eq. (6-19) as the equivalent of a certain percentage of the existing velocity pressure. A pressure loss may be greater or less than the existing velocity pressure. Pressure losses due to frictional resistance in no way affect the velocity pressure, but since these losses are proportional to the square of the velocity, in the same manner as velocity pressures, the velocity pressure can be used as a unit in which to express a pressure loss. Another generally more satisfactory way in which pressure losses are expressed, especially in the case of flow through pipe and duct elbows, is in terms of the additional length of straight run of pipe or duct that would produce the same resistance as the elbow or other form of obstruction.

6-8. Pressure Loss Due to Elbows. When a fluid flows through an elbow or bend in a pipe, the fluid at the outer radius of curvature travels a greater distance than that at the inner radius (or throat) of the bend. A pressure loss, of greater magnitude than in a straight run of pipe, is to be expected because of greater differences in velocities of adjacent particles of fluid. An additional pressure loss also occurs because of the formation of eddy currents which are brought about by sudden changes in direction and velocity of portions of the fluid stream. The increase in frictional resistance is not confined to the elbow or bend itself but extends downstream for the distance necessary for restoration of the normal fully developed flow pattern. The magnitude of the total pressure loss for turbulent flow in an elbow or bend is obviously dependent upon the radius of curvature as well as upon the area of the cross section and upon the roughness of the surface, but for standard elbows and return bends used with wrought iron and steel pipe and with copper tubing the loss may be expressed with fair accuracy in terms of the additional equivalent length of straight pipe or tubing as shown in Table 6-3.

TABLE 6-3. PRESSURE LOSS DUE TO ELBOWS, FOR TURBULENT FLOW

Type of bend	Additional equivalent length, nominal pipe diameters
90-deg elbow	25
45-deg elbow	18
90-deg welded elbow	13
180-deg bend, made by two 90-deg elbows	38
90-deg long-turn elbow	13
180-deg welded return bend	20

Of the two pressure losses which occur in an elbow, the loss due to eddy currents is independent of the surface roughness but is proportional to the velocity pressure, whereas the loss due to viscosity of the fluid is related both to the velocity pressure and to the surface roughness, in the same manner as for turbulent flow in straight pipe.   The combined pressure loss can therefore not be expressed exactly in terms of velocity pressure; nor can the equivalent length of straight pipe truly represent a combination of these two different types of losses.   For flow through elbows in ducts of rectangular cross section a further factor to influence the pressure drop is the aspect ratio.   No satisfactory theoretical treatment has yet been found by which the pressure loss in such elbows can be predicted, but correlation of experimental results[7] has established that here, as in the case with elbows of circular cross section, the equivalent-length concept best serves to express the pressure loss.

Table 6-4 shows the additional length to allow for the pressure drop in 90-deg duct elbows having different radii of curvature and different aspect ratios.   The length as here expressed is in terms of the "width" dimension of the duct in the plane of the bend.

Example 6-4.   Determine the additional equivalent length in feet to allow for the pressure drop in two elbows in a 24- by 6-in. duct.   Elbow A, which is in a horizontal run, changes the direction 90 deg and has a center-line radius of 18 in.   Elbow B changes the direction from horizontal to vertical and has a centerline radius of 9 in.

Solution.   The dimension W (Table 6-4), by definition, is in the plane of the bend and is 24 in. at elbow A and 6 in. at elbow B.

For elbow A, the centerline radius ratio $R/W = {}^{18}\!\!/\!_{24} = 0.75$. The aspect ratio $H/W = {}^{6}\!\!/\!_{24} = 0.25$. L/W (by Table 6-4) = 12. The equivalent length $L = 12W = 12 \times {}^{24}\!\!/\!_{12} = 24$ ft.

For elbow B, the centerline radius ratio $R/W = {}^{9}\!\!/\!_{6} = 1.5$. The aspect ratio $H/W = {}^{24}\!\!/\!_{6} = 4$. L/W (by Table 6-4) = 6.0. The equivalent length $L = 6.0W = 6.0 \times {}^{6}\!\!/\!_{12} = 3$ ft.

TABLE 6-4. EQUIVALENT LENGTH FOR 90-DEG
DUCT ELBOWS
To be added to straight runs of duct measured to the
intersection of their centerlines

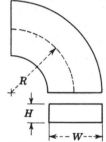

Addition equivalent length, in terms of width, L/W

Aspect ratio, H/W	Centerline radius ratio, R/W					
	0.50	0.75	1.00	1.25	1.50	2.00
0.15	20	10	6.4	4.4	3.2	2.0
0.25	24	12	7.2	4.8	3.5	2.1
0.50	33	15	8.7	5.6	3.9	2.3
1.0	46	19	10.6	6.6	4.4	2.4
2.0	65	26	13	7.8	5.1	2.6
4.0	. . .	35	17	9.5	6.0	2.9
6.0	. . .	42	20	10.6	6.5	3.0

6-9. Computations of Velocity and Velocity Pressure. In
Art. 6-2 expressions were given for velocity pressure as $\rho V^2/2g_c$ lb/sq ft and for velocity head as $V^2/2g$ ft of fluid, both expressions having been developed from the expression for kinetic energy, $V^2/2g_c$ ft-lb/lb$_m$. Numerical values for these expressions may be determined if the velocity V is known or is calculated from knowledge of the volume or weight flow rate as measured by any one of a variety of metering devices.[8] Otherwise, the velocity head may be measured directly as the difference between the total and static heads. This differential head may be indicated

by the height of a column of fluid in a manometer connected to the total-pressure and static-pressure sides of a Pitot tube. The fluid in the manometer may be the same as that which is being metered, or it may be another fluid of known density, whereby the reading of the manometer may be converted into the equivalent height of a column of the fluid under test.

The expression for velocity head, $h_v = V^2/2g$, when transposed, shows that the velocity

$$V = \sqrt{2gh_v} \tag{6-23}$$

where V and g are in the same unit of time, such as V in ft/sec and $g = 32.17$ ft/sec², or as V in ft/hr and $g = 4.17 \times 10^8$ ft/hr²; and h_v is the velocity head in feet of the fluid flowing.

In measurements of the flow of gases the manometer fluid must be a liquid—in order to be visible. In such cases the velocity head is commonly indicated in inches of water, and the conversion of inches of water into equivalent feet of gas changes Eq. (6-23) into the form

$$V = \sqrt{\frac{2g\rho'h'}{12\rho}} = 2.315\sqrt{\frac{\rho'h'}{\rho}} \tag{6-24}$$

where V = velocity, ft/sec.

g = commonly taken, 32.17 ft/sec².

h' = velocity head, in. of water.

ρ = density of the gas, lb_m/cu ft.

ρ' = density of water at the manometer temperature, lb_m/cu ft.

2.315 = a constant, in units of ft½/sec.

Example 6-5.   Air at standard atmospheric pressure and at 70°F flows at the rate of 4,000 cfm from an air chamber which is 5 ft wide by 4 ft high, then through an abrupt entrance to a 50-ft length of 24- by 12-in. galvanized iron duct and is finally discharged into an open room.   Determine (a) the velocity head in the duct and (b) the static and total heads in the air chamber, all expressed in inches of water at a density of 62.3 lb_m/cu ft.

Solution

(a) Air velocity in duct $= \dfrac{Q}{A} = \dfrac{4,000}{60 \times 2} = 33.33$ fps

Velocity head $h_v = \dfrac{V^2}{2g} = \dfrac{33.33 \times 33.33}{2 \times 32.17} = 17.25$ ft of air

or $17.25 \times \dfrac{12\rho}{\rho'} = 17.25 \times \dfrac{12 \times 0.075}{62.3} = 0.249$ in. of water

(b) Equivalent diameter $D_e = \dfrac{4A}{P} = 4 \times \dfrac{2}{6} = 1.33$ ft

$$\text{Relative roughness } \frac{\epsilon}{D} = \frac{0.0005}{1.33} = 0.000375$$

$$N_{Re} = \frac{DV\rho}{\mu} = \frac{1.33 \times 33.33 \times 3600 \times 0.075}{0.0445} = 2.69 \times 10^5$$

$$\text{Friction factor } f \text{ (Fig. 6-4)} = 0.0177$$

Loss of static head in the duct [by Eq. (6-13)] $= f \dfrac{LV^2}{D2g}$

$$= \frac{0.0177 \times 50 \times 33.33 \times 33.33}{1.33 \times 2 \times 32.17} = 11.5 \text{ ft of air}$$

or $\dfrac{11.5 \times 0.075 \times 12}{62.3} = 0.166$ in. of water

Loss of static head at entrance to duct [by Eq. (6-20)] $= 0.44 \times 0.249$
$$= 0.109 \text{ in. of water}$$

The static head in the air chamber is the sum of the loss due to duct friction, the entrance loss, and the static head which is converted into velocity head at the entrance.

$$\text{Static head} = 0.166 + 0.109 + 0.249 = 0.524 \text{ in. of water}$$

Since the cross-sectional area of the air chamber is 10 times that of the duct, the velocity head in the chamber is only $0.01 \times 0.249 = 0.002$ in. of water. The total head, for all practical purposes, may therefore be considered to be equal to the static head.

6-10. Power Required to Move Fluids.

The expenditure of energy for moving a fluid has been observed in Eq. (6-1) in which flow work per unit mass was represented by the term pv, the product of the pressure per unit area and the specific volume. The work done in moving any given mass of fluid against the existing resistance may be expressed as pQ, where p is the pressure per unit area and Q is the volume of the entire mass of fluid. If p is in units of pounds per square foot and Q is the volume in cubic feet, the work is expressed in foot-pounds. Otherwise, the amount of fluid may be denoted by the weight in pounds, and the pressure by the head in feet of the fluid, and the product is similarly in foot-pounds.

Power is defined as the time rate of doing work and is commonly expressed in terms of horsepower—550 ft-lb$_f$/sec, or 33,000 ft-lb$_f$/min. The horsepower input to any device used for moving a fluid, such as a pump or a fan, may be expressed in equation form as

$$Hp = \frac{pQ}{33,000 \times e_m} \tag{6-25}$$

or

$$Hp = \frac{wh}{33,000 \times e_m} \tag{6-26}$$

where p = total pressure difference created by the device, lb_f/sq ft.

Q = volume flow rate of the fluid, cfm.

e_m = mechanical efficiency, expressed as a decimal.

h = total head, ft of fluid.

w = weight flow rate of the fluid, lb/min.

In the foregoing equations the total pressure difference, or total head, is the rise of total pressure from the inlet to the outlet of the device. The total pressure at each location is the algebraic sum of the static pressure and the velocity pressure corresponding to the average velocity. At the outlet of a pump or fan both the static and the velocity pressures are normally above the pressure of the atmosphere, whereas at the inlet the total pressure usually combines a negative static pressure with a positive velocity pressure and the algebraic sum is in most instances less than the pressure of the atmosphere. In the case of a pump the suction and discharge pressures must be measured at the same elevation or must be corrected to the same elevation. In the case of a fan with no inlet duct, the total pressure at the inlet is considered to be zero or the same as the pressure of the surrounding atmosphere and no pressure reading at the inlet is ordinarily taken.

Example 6-6. A centrifugal pump with a 3-in. discharge and a 4-in. inlet opening operates at a mechanical efficiency of 70 per cent when pumping 300 gal of water per minute at 70°F. The discharge and suction pressures are, respectively, 30 psig and 10 in. Hg vacuum, both measured at the same elevation. What is the required horsepower input to the pump?

Solution. The flow may be expressed as

300 × 8.33 = 2,499 lb/min, or 41.65 lb/sec, or as
300 × 231/1,728 = 40.10 cfm, or 0.6684 cfs

Discharge head = 30 × 144/62.27 = 69.38 ft of water
Suction head = −10 × 13.59/12 = −11.33 ft of water
Outlet area = $(\pi/4) \times (\frac{1}{4})^2$ = 0.0491 sq ft
Inlet area = $(\pi/4) \times (\frac{1}{3})^2$ = 0.0872 sq ft

Outlet velocity $= 0.6684/0.0491 = 13.6$ fps
Outlet velocity head $V^2/2g = 13.6^2/(2 \times 32.17) = 2.89$ ft of water
Inlet velocity $= 0.6684/0.0872 = 7.67$ fps
Inlet velocity head $= 7.67^2/(2 \times 32.17) = 0.91$ ft of water
Total head at outlet $= 69.38 + 2.89 = 72.27$ ft of water
Total head at inlet $= -11.33 + 0.91 = -10.42$ ft of water
Total head of pump $= 72.27 - (-10.42) = 82.69$ ft of water,
or total pressure of pump $= 82.69 \times 62.27 = 5,150$ lb/sq ft

$$\text{Horsepower input [by Eq. (6-25)]} = \frac{5,150 \times 40.10}{33,000 \times 0.70} = 8.93 \text{ hp}$$

Otherwise [by Eq. (6-26)],

$$\text{Horsepower input} = \frac{2,499 \times 82.69}{33,000 \times 0.70} = 8.93 \text{ hp}$$

Problems

1. Calculate the critical velocity in feet per minute for hydrogen gas at atmospheric pressure flowing at 32°F in a 6.00-in. (i.d.) tube.

2. Devise a rule for the flow of liquids in pipes. The rule is to show the approximate quantity of liquid in gallons per minute in terms of the square of the nominal pipe diameter in inches, for flow at a velocity of 4 fps.

3. Air at 120°F and 14.7 psia flows through a 6.00-in. (i.d.) tube at the rate of 600 cfm. Find the velocity pressure in inches of water at 70°F.

4. Water at a temperature of 80°F flows through a 3.07-in. (i.d.) galvanized iron pipe at a rate of 300 gpm. Estimate the loss of head in feet of the fluid due to friction for (a) an intermediate section of the pipe 50 ft long, for (b) a 50-ft length starting with an abrupt entrance to the pipe from a tank which has a cross-sectional area ten times that of the pipe, and for (c) a 50-ft length in which the flow is into the tank in the opposite direction to that of part b.

5. Transformer oil having a specific gravity of 0.90, an average temperature of 140°F, and a viscosity as shown in Table 12-2 flows through a 3.00-in. (i.d.) smooth-drawn tube at a velocity of 6 fps. Estimate the pressure loss in lb$_f$/sq ft for a 40-ft length.

6. Repeat Prob. 5, but for nonisothermal flow where the average tube-wall temperature is 110°F.

7. Air at atmospheric pressure and at an average temperature of 100°F flows through a 50-ft horizontal length of galvanized iron duct 12 in. wide and 6 in. deep in which there is an elbow having a centerline radius of 9 in. (a) What is the equivalent length of the duct for calculation of the pressure drop? (b) What is the equivalent length if the elbow changes the direction of flow from horizontal to vertical?

8. What is the equivalent length of ¾-in. (i.d.) tubing in a heat exchanger in which there are 10 lengths of 5 ft each, made into a continuous circuit by the use of nine 180-deg welded return bends?

9. A centrifugal fan operates at a mechanical efficiency of 70 per cent when discharging 8,000 cfm of air at a density of 0.075 lb$_m$/cu ft against a static resistance of 1.50 in. of water and with a suction at the fan inlet of 0.30 in. of water. The outlet area of the fan is 4.0 sq ft, and the inlet area is 5.0 sq ft. Assume the density of water at the temperature of the manometer to be 62.3 lb_m/cu ft and g to be 32.17 ft/sec². Determine the horsepower input to the fan.

REFERENCES

1. Reynolds, O.: *Phil. Trans. Royal Soc.* (*London*), **A, 174,** pt. III, p. 935 (1883), or "Scientific Papers," Cambridge University Press, vol. II, pp. 51–105 (1900–1903).
2. Moody, L. F.: Friction Factors for Pipe Flow, *Trans. ASME,* **66,** 671–694 (1944).
3. Sieder, E. N., and G. E. Tate: *Ind. Eng. Chem.,* **28,** 1429 (1936).
4. Huebscher, R. G.: Friction Equivalents for Round, Square and Rectangular Ducts, *Trans. ASHVE,* **54,** 101–108 (1948).
5. Schlichting, H.: "Boundary Layer Theory," McGraw-Hill Book Company, Inc., New York, 1955.
6. Hughes, H. C., and A. T. Safford: "Hydraulics," The Macmillan Company, New York, 1926.
7. Lochlin, D. W.: Energy Losses in 90-degree Duct Elbows, *Trans. ASHVE,* **56,** 479–502 (1950).
8. Rhodes, T. J.: "Industrial Instruments for Measurement and Control," pp. 186–337, McGraw-Hill Book Company, Inc., New York, 1941.

CHAPTER 7

FORCED CONVECTION

7-1. General. The transfer of heat to or from a fluid (liquid or gas) flowing over a surface of a hotter or colder body is by the process known as convection. Convection is termed *free*, or *natural*, when circulation is produced because of differences in density resulting from temperature changes, as when water is heated in a boiler. When circulation is made positive by some mechanical means, such as a pump or fan, convection is termed *forced convection*.

The principle of forced convection is utilized, either wholly or in part, in a large proportion of the examples of heat transfer. In forced warm-air systems, heat is removed from hot surfaces of the furnace by forced convection and is transferred to the rooms and to their occupants by the same process. In refrigeration and in air conditioning, heat is transferred largely by forced convection. An automobile radiator* receives heat from the water jackets of the motor cylinders by forced convection and delivers it by the same process to the air that is drawn over the radiator surfaces. Forced convection plays the chief role in the transfer of heat in air heaters, air- and water-cooled condensers, in driers, aircraft heaters, regenerators, in coils used in the refining of oils, in electric generators and motors, in large electric transformers, and in many forms of heat exchangers wherever heat is added to or removed from air, water, brines, or various other liquids or gases.

7-2. The Film Theory. From the work of investigators in the fields both of fluid flow and of heat transfer, the concept has developed that when a fluid flows over a surface (whether in

---

* The name "radiator" for such equipment is obviously a misnomer, since radiation here is practically absent; the name can be accounted for only by the fact that it was applied before the time when much thought was given to the processes of heat transfer.

streamline or in turbulent flow), a stagnant film adheres to the surface and acts as a heat insulator.   Experiments[1] have shown the actual existence of such a film.   No attempt is ordinarily made to measure the thickness of the film, which may be immeasurably thin or several hundredths of an inch in thickness, but in the study of heat transfer it may be visualized as a barrier to the flow of heat, a barrier that adheres to the surface but is partially wiped off and accordingly reduced in effectiveness as the velocity of the fluid is increased.

In the process of heat transfer by convection, heat is transferred really by conduction through the stagnant portion of the film and is then transferred to the moving particles and carried away by convection currents into the main portion of the fluid stream.   In the study of heat transfer by convection, however, the concept of the film as being completely stagnant appears to be the least complicated and generally most satisfactory approach to the problem.   Heat is presumed to be conducted through this film in proportion to the size and shape of the surface, the specific heat and conductivity of the fluid, the difference in temperatures on the two sides of the film, and its thickness which, although not measured, has been found to be dependent upon its viscosity and density and upon the velocity of the fluid stream.

7-3. Surface, or Film, Conductance.   The flow of heat through the fluid film that is assumed to adhere to the surface of any solid in contact with a fluid may be expressed as

$$q = hA(t_1 - t_2) \qquad (7\text{-}1)$$

where q = rate of heat flow, usually expressed in Btu/hr.

h = surface coefficient (or film conductance)* or the rate of heat transfer per unit area per degree of temperature change, Btu/(hr)(sq ft)(°F).

A = area of the surface, sq ft.

$t_1 - t_2$ = difference in temperature on the two sides of the film, °F.

or for unit area

$$\frac{q}{A} = h(t_1 - t_2) \qquad (7\text{-}2)$$

---

* The terms *surface coefficient* and *film conductance* are generally used synonymously.   The latter term, however, is hardly appropriate in cases where radiation is involved.

Equations (7-1) and (7-2) may be applied for the purpose of establishing the rate of heat flow q wherever it is feasible to measure the temperatures t_1 and t_2; otherwise, where the rate of heat flow is known, calculation of the value of h makes it possible to determine the temperature difference $t_1 - t_2$.

Heat may be transferred at the surface by convection alone, in which case the surface coefficient will be designated as h_c; or the transfer may be by radiation, in which case the surface coefficient will be designated as h_r. Where both convection and radiation occur, the coefficient h will equal $h_c + h_r$.

This chapter deals with the surface coefficient of forced convection and in specific cases with the surface coefficient of both convection and radiation. Following chapters will deal with surface coefficients of free convection, of boiling liquids, and of condensing vapors.

7-4. Surface Coefficient of Forced Convection. In the development of a mathematical expression for the surface coefficient of forced convection, where the flow is assumed to be turbulent, three different means of approach may be employed. The first is by a mathematical analysis of fluid flow, translated into thermal units through an analogy between fluid friction and heat transfer. The second is by application of the principles of dimensional analysis together with the introduction of numerical constants derived from experimental data. The third, which has less general application, is by the representation of experimental data by purely empirical formulas.

The first means of approach had its beginning with the work of Osborne Reynolds[22] in 1874 through his observation that in geometrically similar systems of piping, the transfer of heat by convection is definitely related to the fluid friction. In more recent years, Reynolds' analogy has been progressively extended by Taylor,[2] Prandtl,[3] von Kármán,[4] Bakhmeteff,[5] and Boelter, Martinelli, and Jonassen[6] in the attempt to develop an expression that will show close agreement with experimental results for all fluids of known physical properties throughout the widest ranges of temperatures and velocities.

The results of these endeavors are a number of expressions that agree quite closely with one another and with experimental results in certain ranges of physical properties and to a lesser degree in other ranges. More specifically, in the case of gases where, for all temperatures, the product of the specific heat and absolute viscosity is usually very similar in numerical value to the thermal conductivity ($c_p\mu = k$), the agreement is good; but where $c_p\mu/k$ is of the order of 10 or more, as in the case for viscous liquids, the agreement is not so good.

The expressions devised by rigid mathematical analysis of the analogy between heat transfer and fluid friction are quite involved, and their derivations are so much more complicated than the development of an expression by the method of dimensional analysis that the latter procedure generally has met with greater favor.   In the matter of accuracy of results, as will be shown later, there is little choice between the expressions developed by the two different means; accordingly, the simpler method of dimensional analysis will be employed in the following development of an expression.

In order to develop a mathematical expression for the surface coefficient of forced convection where the flow is turbulent, it is necessary first to establish the factors that may influence the conductivity and the thickness of the stagnant film that is assumed to adhere to the surfaces over which convection currents pass.   According to the laws of fluid flow and of heat transfer by conduction, it is reasonable to expect the film conductance at any point to be influenced by the size and shape of the surface; by the thermal conductivity, specific heat, viscosity, and density of the fluid at the mean temperature of the film; and by the velocity of the fluid stream.

If the size and shape of the stream are expressed in terms of the diameter, the surface coefficient of convection may be stated in equation form as

$$h_c = \text{function of } D,\ V,\ \mu,\ \rho,\ c_p,\ k$$

where h_c = surface coefficient of convection.

D = diameter of the fluid stream.

V = velocity.

μ = absolute viscosity.

ρ = density.

c_p = specific heat at constant pressure.

k = thermal conductivity.

A mathematical expression for h_c will now be developed by the method of dimensional analysis.   According to the π theorem the equation may be expressed in the form

$$\pi = h_c{}^a D^b V^d \mu^e \rho^f c_p{}^g k^m$$

All the variables may be expressed in the fundamental terms of length L, time T, mass M, and temperature change θ, provided consistent units are used, such as L in feet, T in hours, M in pounds mass, and θ in degrees. Expressed as dimensional formulas

$$D = L \qquad\qquad c_p = \frac{H}{M\theta} = \frac{FL}{M\theta} = \frac{L^2}{T^2\theta}$$

$$V = \frac{L}{T} \qquad\qquad k = \frac{H}{TL\theta} = \frac{FL}{TL\theta} = \frac{ML}{T^3\theta}$$

$$\mu = \frac{FT}{L^2} = \frac{M}{LT} \qquad h_c = \frac{H}{TL^2\theta} = \frac{FL}{TL^2\theta} = \frac{M}{T^3\theta}$$

$$\rho = \frac{M}{L^3}$$

Expressed by dimensions, the equation becomes

$$\pi = \left(\frac{M}{T^3\theta}\right)^a L^b \left(\frac{L}{T}\right)^d \left(\frac{M}{LT}\right)^e \left(\frac{M}{L^3}\right)^f \left(\frac{L^2}{T^2\theta}\right)^g \left(\frac{ML}{T^3\theta}\right)^m$$

or $\qquad \pi = M^{a+e+f+m} L^{b+d-e-3f+2g+m} T^{-3a-d-e-2g-3m} \theta^{-a-g-m}$

from which the following simultaneous equations are obtained:

$$a + e + f + m = 0$$
$$b + d - e - 3f + 2g + m = 0$$
$$-3a - d - e - 2g - 3m = 0$$
$$-a - g - m = 0$$

Since there are seven quantities expressed in terms of four fundamental dimensions, there will be three dimensionless groups, or three π terms, in the result.

Since an expression for h_c in terms of the other variables is desired, it is logical to let $a = 1$. Arbitrarily assigned values of $d = 0$ and $e = 0$ lead to the result $b = 1$, $f = 0$, $g = 0$, and $m = -1$. Therefore

$$\pi_1 = \frac{h_c D}{k}$$

To find the second dimensionless group, π_2, in which it would be undesirable to have h_c appear again, the assumption should be made that $a = 0$. Arbitrarily assigned values of $b = 1$

and $g = 0$ lead to the result $d = 1$, $e = -1$, $f = 1$, and $m = 0$. Therefore

$$\pi_2 = \frac{DV\rho}{\mu}$$

To find the third dimensionless group, π_3, the assumption may again be made that $a = 0$; then arbitrarily assigned values of $g = 1$ and $f = 0$ lead to the result $b = 0$, $d = 0$, $e = 1$, and $m = -1$. Therefore

$$\pi_3 = \frac{c_p\mu}{k}$$

By the π theorem, then,

$$\phi\left(\frac{h_cD}{k}, \frac{DV\rho}{\mu}, \frac{c_p\mu}{k}\right) = 0$$

and a solution of this equation is

$$\frac{h_cD}{k} = f\left(\frac{DV\rho}{\mu}, \frac{c_p\mu}{k}\right)$$

or

$$\frac{h_cD}{k} = f_1\left(\frac{DV\rho}{\mu}\right)f_2\left(\frac{c_p\mu}{k}\right)$$

The form in which the functions f_1 and f_2 of $DV\rho/\mu$ and $c_p\mu/k$ can readily be expressed has been determined by plotting the results of numerous tests. For most practical applications, these results are well expressed by the equation

$$\frac{h_cD}{k} = C\left(\frac{DV\rho}{\mu}\right)^b\left(\frac{c_p\mu}{k}\right)^d \tag{7-3}$$

This equation is known as Nusselt's[7] expression, and the three fractions within the brackets are known as follows:

$\dfrac{h_cD}{k}$ Nusselt number, or modulus, or the boundary modulus

$\dfrac{DV\rho}{\mu}$ Reynolds number, or modulus

$\dfrac{c_p\mu}{k}$ Prandtl number, or modulus

$\dfrac{N_{Nu}}{N_{Re} \times N_{Pr}} = \dfrac{h_c}{Vc_p\rho}$ is known as the Stanton number or modulus.

When expressed in consistent units, the numerator of each of the three fractions has the same dimensions as the denominator; so all three numbers are seen to be dimensionless.

Any physical significance that may be attached to the Nusselt, Reynolds, and Prandtl numbers is unimportant, but extreme care must be exercised in the calculation of these numbers to see that consistent units are used in all terms. For example, if the diameter of the fluid stream is expressed in feet, then the velocity must be expressed in feet per unit of time, and not in inches, miles, meters, or any other unit. Thermal conductivity likewise should be on the basis of feet. Again, if the density ρ, which is commonly defined as the mass per unit volume, is expressed in pounds mass per cubic foot, then the same unit, the pound mass, must be used in the expressions for viscosity and for specific heat.

Unfortunately, tabulated values of viscosity appear in the literature in at least six different units of mass or weight, distance, and time; but the unit in the English system that is most consistent and convenient for calculations in heat-transfer problems is the pound mass per foot-hour.* In this unit viscosity is shown in tables and graphs included in the appendix, with the exception of Figs. A-6a and A-6b where the unit is the centipoise. Conversion factors for viscosity units are shown on page 84.

7-5. Verification of Nusselt's Equation Experimentally. The work of a goodly number of experimenters has shown that the film conductance for a liquid being heated or cooled is related to all the factors assumed in the development of Nusselt's equation, and no evidence has been presented that any factors have been overlooked. For the heating and cooling of *gases*, where the Prandtl number varies only from approximately 0.65 to 0.84 for the common gases and is approximately 1.20 for steam at 300°F and 1 atm pressure, there is little opportunity for determining by test whether or not this term should be included in the equation. Since there is no experimental evidence to the contrary, it may logically be assumed that Nusselt's equation is generally applicable to problems of film conductance for all fluids not only in the liquid but also in the gaseous state.

---

* For calculations in hydraulics involving the Reynolds number, the more convenient unit for viscosity is the pound mass per foot-*second*, since, in this field, velocities are usually expressed in feet per *second* and no other term in the Reynolds number involves the time element.

Numerical values of C, b, and d, for substitution in Eq. (7-3), have been determined by a number of experimenters for a variety of fluids tested under a number of different conditions. The values for flow inside pipes are not the same as for flow across the outside; and different values of C and d have been found for heating and for cooling liquids in pipes. The number and arrangement of pipes or tubes in a tube bank also affect the value of the constant C.

7-6. Film Conductance of Fluids in Pipes and Tubes. Morris and Whitman[8] investigated the film conductance for both the heating and the cooling of water, gas oil, straw oil, and light motor oil inside a pipe and found the same relationships among the Nusselt, Reynolds, and Prandtl numbers for the various liquids. In their tests the liquid was heated or cooled while flowing through a ½-in. standard steel pipe (i.d. 0.62 in.) at velocities of 1 to 20 ft/sec, with heat-transfer rates of 10 to 700 Btu/(hr)(sq ft)(°F). Viscosity ranged from 1.2 to 133 lb_m/(ft)(hr). The temperature in the heating tests ranged from 83 to 169°F and in the cooling tests from 115 to 512°F. Computations of Reynolds number for these ranges will show that not all the tests were in the range of turbulent flow; but if the results of tests with turbulent flow are expressed in the form of Nusselt's equation, the values of C, b, and d, corresponding to different values of the Reynolds and Prandtl numbers, are fairly constant. Since C, b, and d, however, were found to show some variation, Morris and Whitman preferred to express their results in the form

$$N_{Nu} = \phi(N_{Re})\psi(N_{Pr}) \qquad (7-4)$$

where N_{Nu}, N_{Re}, and N_{Pr} = Nusselt, Reynolds, and Prandtl numbers and ϕ and ψ = variable functions of N_{Re} and N_{Pr}, respectively, which they showed graphically.

Others who have investigated the relationships among the Nusselt, Reynolds, and Prandtl numbers as shown by plots of test data have considered the variations in the values of the constant C and the exponents b and d to be within the range of experimental accuracy and have preferred to assign fixed values to these terms for use in Eq. (7-3). Agreement does not exist among the various experimenters, however, as to the proper or most accurate values to be assigned; nor has it been established whether the variations in recommended values are due to differences in techniques in the measurement of heat quantities and temperatures, to differences in test conditions, to differences in the temperatures at which the physical properties of the fluids have been evaluated, to experimental errors, or to factors which are still unknown.

From a review of the work of various experimenters, McAdams[9] has concluded that a fair correlation of their results for the heating and cooling of various fluids in turbulent flow in horizontal tubes is shown by the equation

$$h_c = 0.023 \frac{k}{D} \left(\frac{DV\rho}{\mu}\right)^{0.8} \left(\frac{c_p\mu}{k}\right)^{0.4} \qquad \text{Btu/(hr)(sq ft)(°F)} \qquad (7\text{-}5)$$

This equation applies where the Reynolds number is within the range of 10,000 to 120,000, the Prandtl number is between 0.7 and 120, the length of the tube is at least 60 diameters, and the difference in temperature on the two sides of the film is not large. The physical properties of the fluid are to be evaluated at the temperature of the main body of the stream (often termed the bulk temperature*). Modifications for making this equation applicable to conditions other than those which have been specified will be presented later in this chapter.

In Eq. (7-5) and in all following equations which show the relationships of the Nusselt, Reynolds, and Prandtl numbers, consistent units must be used. In applications of these expressions in the text, all dimensions, unless otherwise noted, will be expressed in feet and the units of time and mass will be hours and pounds, respectively. The surface coefficient h_c may be a point value or an average value; i.e., it may apply to any point along a tube where the temperatures on both sides of the film are known or it may be an average value applying to the entire inside surface of the tube. In the latter case the average temperature difference is usually considered to be the difference between the arithmetic mean of the surface temperatures and of the fluid temperatures at the two ends of the tube.

It may now be recalled that Nusselt's expression, Eq. (7-3), has been developed upon the hypothesis that heat in being transferred from a surface to a turbulent stream passes by

---

* This temperature is also called the mixing-cup temperature—the temperature that would result if the stream of fluid at any point along its path were completely mixed as in a cup.

conduction through a stagnant film that adheres to the surface. No account was taken of the intermediate layer of eddy currents described in Art. 6-3. In the more rigid mathematical analysis made by Prandtl and von Kármán, applied to turbulent flow in pipes, the presence of the intermediate layer between the turbulent core of the stream and the stagnant film is not overlooked.

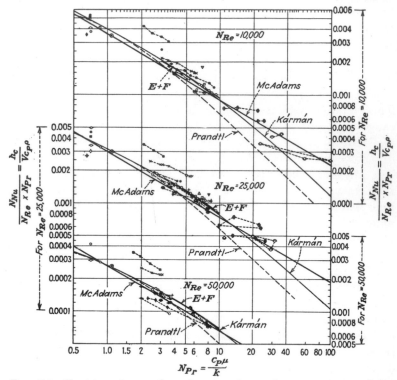

FIG. 7-1. Heat-transfer numbers for pipes by various equations and by experimental data of Eagle and Ferguson.

These analyses and related studies of fluid friction show that the thermal resistance of the turbulent core is related to the Reynolds number but is independent of the Prandtl number, whereas the thermal resistance of the laminar sublayer, for a constant Reynolds number, is proportional to the Prandtl number. The thermal resistance of the intermediate layer lies between the two other values.

Figure 7-1 shows a curve of McAdams' equation superimposed on a graph used by von Kármán[10] to show the comparison of Prandtl's and von Kármán's equations with experimental data

obtained by Eagle and Ferguson.[11]   The equations plotted in
this figure are the McAdams equation:

$$N_{Nu} = 0.023 N_{Re}^{0.8} N_{Pr}^{0.4} \qquad (7\text{-}6)$$

the Prandtl equation:

$$N_{Nu} = \frac{0.04 N_{Re}^{3/4} N_{Pr}}{1 + 1.74 N_{Re}^{-1/8}(N_{Pr} - 1)} \qquad (7\text{-}7)$$

and the von Kármán equation:

$$N_{Nu} = \frac{0.04 N_{Re}^{3/4} N_{Pr}}{1 + N_{Re}^{-1/8}\{N_{Pr} - 1 + \ln[1 + 5/6(N_{Pr} - 1)]\}} \qquad (7\text{-}8)$$

It may be noted that in the range of the Prandtl number N_{Pr}
or $c_p \mu/k$ up to 5, for all three values of the Reynolds number,
viz., 10,000, 25,000, and 50,000, there is reasonably close agreement between all the equations and the experimental results.
Even at values of N_{Pr} equal to 20, McAdams' and von Kármán's
equations are in fair agreement.   At higher values of N_{Pr} in
the region where experimental data are unfortunately very
limited, the graphs diverge progressively with decreasing values
of the Reynolds number.   From this comparison it is difficult
to conclude which of the three equations best represents correct
experimental results.   This is especially the case in the range
of higher values of N_{Pr}, which apply to viscous fluids such as the
heavier oils.   At least until more conclusive results of tests are
obtained in the range of the higher values of the Prandtl number,
the general use of the least complicated equation, McAdams'
equation, appears justifiable.

For low values of the Prandtl number ($N_{Pr} < 0.5$) the Mc-
Adams, Prandtl, and von Kármán equations predict values of
Nusselt numbers that are greatly in error when compared to
experimental data.   This is of particular significance since liquid
metals, which are useful heat-transfer fluids in atomic reactors,
have values of N_{Pr} less than 0.1.   In an attempt to account for
this large error, Martinelli[12] derived an equation based on the
following assumptions:

1. Heat and momentum are transferred by molecular action
alone in the laminar sublayer (heat by thermal conduction and
momentum by virtue of the viscosity).

2. Heat and momentum are transferred by molecular action
and by eddy diffusion in the transition layer.

3. Heat is transferred by thermal conduction and eddy diffusion in the turbulent core, and momentum is transferred by eddy diffusion alone.

4. The Nikuradse[13] velocity distribution* is valid.

With these assumptions Martinelli's equation for fully established turbulent flow in an isothermal tube is

$$N_{Nu} = \frac{0.04 N_{Pr}(N_{Re})^{7/8}}{N_{Pr} + \ln(1 + 5N_{Pr})_J^{'} + \dfrac{1}{2(1 + \beta)} \ln \dfrac{(N_{Re})^{7/8}}{300}} \qquad (7\text{-}9)$$

In this equation β is a factor which depends on the ratio of molecular conduction to the eddy-diffusion heat transfer in the turbulent core. Some values of β from Martinelli's curves are as follows:

N_{Re}	$N_{Re} \times N_{Pr}$			
	10^2	10^3	10^4	10^5
10^4	0.175	0.67	0.92	1.00
10^5	0.095	0.45	0.82	1.00
10^6	0.050	0.28	0.65	0.95

* From Nikuradse's measurements of velocity distributions close to a wall over which a fluid flows in the turbulent region, von Kármán developed generalized expressions for the thicknesses of the laminar sublayer and the buffer zone. He also was able to write equations for the velocity distributions in these regions.

According to von Kármán, the velocity at any point in the laminar sublayer, which extends over the region $0 < y < 5\phi\nu$, is

$$v_L = \frac{y}{\nu\phi^2}$$

and the velocity at any point in the buffer zone, which extends over the region $5\phi\nu < y < 30\phi\nu$, is

$$v_b = \frac{5}{\phi} \ln \frac{y}{\nu\phi} - \frac{3.05}{\phi}$$

where y = distance from the wall.

ϕ = function of density and wall shearing stress.

ν = kinematic viscosity.

v_L = velocity at any point in the laminar sublayer.

v_b = velocity at any point in the buffer zone.

It will be noted from the above that the thickness of the buffer zone is five times the thickness of the laminar sublayer and that the turbulent core is reached when $y = 30\phi\nu$.

Lyon[14] developed an approximation of the Martinelli equation for values of $N_{Pr} < 0.1$. This equation, which gives values within 10 per cent of the Martinelli values, is

$$N_{Nu} = 7 + 0.025(N_{Re} \times N_{Pr})^{0.8} \qquad (7\text{-}10)$$

While Eqs. (7-9) and (7-10) give somewhat more accurate results for extremely low Prandtl numbers than Eqs. (7-6), (7-7) and (7-8), they nevertheless predict values of the Nusselt number which are still somewhat high.

All calculations of the surface coefficient made by applying such equations as (7-6), (7-7), (7-8), (7-9), and (7-10) involve an evaluation of the physical properties of the fluid at some definite temperature. At low rates of heat transfer, the difference between the bulk temperature and that of the surface with which the fluid is in contact is insignificant and the physical properties are usually evaluated at the bulk temperature. At higher rates of heat transfer, the temperature which is usually selected is either the bulk temperature or a mean of the bulk and the surface temperatures. Uniform procedure in the selection of the temperature has not been attained.

Variations in temperature across the diameter of a pipe have been found to occur in much the same manner as the variations in velocity; thus, in turbulent flow there is little variation in temperature across the portion of the stream that is moving rapidly, but a greater variation in the portions near the surface that are moving at a slower rate or are practically stagnant. At the higher rates of heat transfer, not only is there a greater difference between the bulk and the film temperatures, but for a given bulk temperature the film temperature is appreciably higher during heating of the fluid than during cooling. In keeping with the film theory, the significant temperature is that which corresponds to the average physical properties of the stagnant film. This temperature cannot be measured directly and can only be estimated and is usually assumed to be the mean of the bulk and surface temperatures.

Where the surface temperature is not determined by measurement, it can be found for any known rate of heat flow by the trial-and-error method applied to Eq. (7-2), which states that

$$\frac{q}{A} = h(t_1 - t_2) \qquad \text{Btu/(hr)(sq ft)}$$

Where the rate of heat flow is not known but the temperatures of the heating and cooling mediums are known, the surface temperatures can be estimated. The changes in temperature across the various barriers interposed between the hot and cold mediums will be in proportion to the thermal resistances of the barriers. Calculation of surface temperatures by this procedure will be found in Chap. 11, Ex. 11-3.

Examination of the values of c_p, k, μ, and ρ for various fluids at different temperatures will show that in the case of liquids, especially the viscous liquids, the property that is most affected by change in temperature is the viscosity μ; and in the case of gases, significant changes occur in all four properties, although the composite effect of all factors upon the surface conductance has been found to be relatively small.

On the basis of the foregoing observation and as a result of tests with viscous oils, Sieder and Tate[15] concluded that the surface conductances for both the heating and the cooling of viscous liquids in turbulent flow is well expressed by

$$h_c = 0.027 \frac{k}{D} \left(\frac{DV\rho}{\mu}\right)^{0.8} \left(\frac{c_p\mu}{k}\right)^{\frac{1}{3}} \left(\frac{\mu}{\mu_s}\right)^{0.14} \qquad \text{Btu/(hr)(sq ft)(°F)}$$

$$(7\text{-}11)$$

where all physical properties are evaluated at the bulk temperature, except μ_s, which is the viscosity at the surface temperature.

In tests of the heating of air with high temperature differences across the film, a number of investigators[16] have found best agreement between test results and Eq. (7-5) when the physical properties of the air were evaluated at the mean of the bulk and the surface temperatures.

7-7. Recommended Procedure in Calculating the Surface Coefficient. In the light of the foregoing observations and test data, the following procedure is recommended in calculating the surface coefficient for turbulent flow in pipes and tubes at least 40 diameters in length. Apply Eq. (7-5) for both the heating and the cooling of all fluids that are not more viscous than water, with the physical properties evaluated at the bulk temperature, provided the temperature difference on the two sides of the film is not more than 10°F in the case of a liquid and not more than 100°F in the case of a gas. For larger temperature differences, apply the same equation but evaluate the physical properties at

the arithmetic mean of the bulk and surface temperatures.*
In the case of a liquid more viscous than water, apply Eq. (7-11).
For turbulent flow in short pipes and tubes, apply correction
factors as shown in Art. 7-11.

Equation (7-5) has been specified as applicable where the
Reynolds number is within the range of 10,000 to 120,000 and
the Prandtl number is between 0.7 and 120. For Reynolds
numbers below 10,000 but above 3,100, this equation and Eq.
(7-11) may be applied satisfactorily provided the Prandtl number
does not exceed 10.

For Prandtl numbers below 0.7, Eq. (7-9) or Eq. (7-10) should
be used.

Although all the foregoing recommended equations are based
upon the results of tests with horizontal tubes, apparently they
can be applied satisfactorily to flow in vertical tubes and pipes
except in the case of liquids near the boiling point, where the
position of the tube may influence the liberation of bubbles of
vapor from the surface. Experimental results for turbulent
flow in vertical tubes are very limited; but where tubes of any
considerable length are used in heat-exchange applications they
are customarily in a horizontal position.

7-8. The Equivalent Diameter.   The equivalent diameter D_e
of noncircular conduits, as applied in calculations of pressure loss
in turbulent flow, was defined in Art. 6-5 as four times the cross-
sectional area divided by the perimeter.   In calculations of surface conductance of forced convection for turbulent flow, this
same definition applies except that the perimeter includes only
that portion through which heat is transferred.   Thus, in the
case of a double-pipe heat exchanger, in calculations of the
coefficient of forced convection for the outer surface of the inner
pipe or tube, $D_e = (D_2{}^2 - D_1{}^2)/D_1$, where D_1 is the outside
diameter of the inner pipe and D_2 is the inside diameter of the

---

* Evaluation of the density of the fluid may be made at the bulk temperature instead of at the film temperature, in keeping with the contention that
the Reynolds number, often expressed as DG/μ, is primarily related to the
flow rate G, which depends upon the density of the main body of the stream
and not upon the density of the film.   This procedure is at variance with
the concept of convection as a process primarily of conduction through a
stagnant film having physical properties dependent upon its temperature.
Which of the practices produces the better correlation with test results has
not been conclusively established.

outer pipe.   If the fluid flowing in the annular space between the pipes is a gas, radiant-heat exchange will occur between the surfaces of the two pipes in addition to the transfer of heat by convection at the outer surface of the enclosed pipe.

7-9. Effect of Surface Condition.   Equations (7-5) and (7-11) apply to the turbulent flow of fluids in commercial wrought-iron and steel pipes and in metal tubes in a clean condition.   Any fouling of the inside surface of the pipe or tube with grease, rust, scale, or even gas bubbles will materially reduce the film conductance.   Even the presence of a light smoke film, with gas flowing through a brass tube, has been found by Parsons and Harper[17] to cause a 10 per cent reduction in heat transfer.   Likewise, the thorough removal of the oxide film normally found on brass pipe, by means of polishing, produced a 17 per cent increase in heat transfer.   The effects of similar changes in the condition of the surface should be even greater in the case of heat transfer for water in pipes, since the resistance of the water film itself is relatively less than that of the gas film.   In tests of ammonia condensers the normal accumulation of foreign matter on the surfaces during an operating period as short as 16 hr has been found to have decreased the over-all heat transfer in a marked degree.[18]

Any roughing of the surface causes an increase in turbulence, and consequently an increase in film conductance.   This is especially true for flow in small pipes where a given degree of roughness is proportionately greater than in large pipes.   Intentional roughening of a surface in order to increase heat transfer can have only limited practical application, however, since the increase in heat transfer due to roughness is accompanied by the disadvantage of a still greater increase in frictional resistance to flow.

7-10. Character of Flow in Pipes and Tubes.   Expressions for film conductance for pipes and tubes generally apply to positions along the length where the flow pattern is fully developed.   The effects upon the flow pattern caused by abrupt entrances and by sudden enlargements were discussed in Chap. 6; and in calculations of film conductance, allowance for such effects must be observed.

In the region of fully developed flow, the film conductance for turbulent flow is related both to the viscous shear between adjacent layers flowing at different velocities within the laminar film and to the heat which is transported by innumerable eddy currents within the turbulent core. In this region the amount of the viscous shear and the degree of turbulence are fairly well established; but where the flow pattern is changed, as in the cases of abrupt entrances and sudden enlargements, progressive changes of uncertain magnitude occur both in the viscous shear and in the turbulence. In the region where the stream expands, as it does throughout the major portion of the inlet length and downstream of a sudden enlargement, turbulence is greatly increased and higher values of film conductance are to be expected.

Morris and Whitman,[8] in their tests to which reference has already been made, took the precaution of using "calming sections" 20 pipe diameters in length at each end of the test section, realizing the possible effect on heat transfer due to disturbance at the pipe entrance. Any pipe fitting, such as an elbow close to the test section, that would produce swirl or otherwise increase turbulence may be expected to increase the value of h_c.

The insertion of a guide vane into an air duct, lengthwise of the duct but so twisted as to produce a corkscrew type of air flow, has been found to effect a 50 per cent increase in film conductance, but at the disadvantage of a 100 per cent increase in frictional resistance.

7-11. Effect of Abrupt Entrance to Pipes and Tubes. The magnitude of the effect of an abrupt entrance to a pipe or tube upon the film conductance has been a matter of uncertainty. Investigations by G. E. Luke[19] indicate that where the pipe or

TABLE 7-1

Length of tube in terms of diameters..........................	1	5	10	15	20	25	30	35
Shortness factor................	1.76	1.54	1.34	1.22	1.14	1.09	1.05	1.01

tube length is less than 40 diameters, an allowance for the increased film conductance should be observed. Based upon Luke's work, A. D. Moore[20] recommends that for determining the film conductance of short tubes the calculated values for long tubes be multiplied by the factors shown in Table 7-1. These factors are applied equally to tubes with sharp and with curved entrances; for although the curved entrance produces no vena contracta within the tube, it does cause the velocity of the fluid near the tube wall to be nearly equal to the average velocity of the stream, thereby increasing the viscous shear within the laminar boundary layer and consequently increasing the film conductance.

7-12. Application of Nusselt's Expression. The following example will illustrate the application of Nusselt's expression in the form of Eq. (7-5).

Example 7-1. Determine the surface coefficient due to convection, in Btu/(hr)(sq ft of inside tube surface)(°F of temperature difference), when 15,000 lb of water at an average stream temperature of 100°F is being heated per hour in a clean smooth horizontal tube of 2-in. diameter. Assume the difference between stream and film temperatures to be small.

Solution. For water at 100°F (from Table A-1)

$$k = 0.364 \text{ Btu/(hr)(sq ft)(°F/ft)}$$
$$\rho = 61.99 \text{ lb}_m/\text{cu ft}$$
$$\mu = 1.65 \text{ lb}_m/\text{(ft)(hr)}$$
$$c_p = 0.997 \text{ Btu/(lb}_m)(°F)$$
$$D = 2.00 \text{ in.}/12 = 0.1667 \text{ ft}$$

Inside transverse area of tube = 0.0218 sq ft. Average velocity of water,

$$V = \frac{15,000}{61.99 \times 0.0218} = 11,100 \text{ ft/hr}$$
$$N_{Re} = \frac{DV\rho}{\mu} = \frac{0.1667 \times 11,100 \times 61.99}{1.65} = 69,500$$

(which is well above the critical N_{Re}). By Eq. (7-5)

$$h_c = 0.023 \frac{k}{D} (N_{Re})^{0.8} \left(\frac{c_p\mu}{k}\right)^{0.4}$$
$$= 0.023 \times \frac{0.364}{0.1667} (69,500)^{0.8} \left(\frac{0.997 \times 1.65}{0.364}\right)^{0.4}$$
$$= 0.0502 \times 7470 \times 1.828 = 686 \text{ Btu/(hr)(sq ft)(°F)}$$

Equation (7-5) is again applied in the following example.

Example 7-2. Determine the surface coefficient of convection, in Btu/(hr)(sq ft of inside pipe surface)(°F of temperature difference), when superheated steam at 250 psi abs pressure and 600°F flows through an extra heavy 8-in. pipe at a velocity of 10,000 fpm and is being cooled, although not to the saturation temperature.

(Actual pipe diameter = 7.625 in.)

Solution

k (Fig. A-4)   = 0.0248 Btu/(hr)(sq ft)(°F/ft)
ρ (Keenan and Keyes)   = 1/2.427 = 0.412 lb$_m$/cu ft
μ (Fig. A-8)   = 0.063 lb$_m$/(ft)(hr)
c_p (Keenan and Keyes)   = 0.535 Btu/(lb$_m$)(°F)
D = 7.625/12   = 0.635 ft

By Eq. (7-5)

$$h_c = 0.023 \frac{k}{D} \left(\frac{DV\rho}{\mu}\right)^{0.8} \left(\frac{c_p\mu}{k}\right)^{0.4}$$

$$= 0.023 \times \frac{0.0248}{0.635} \left(\frac{0.635 \times 600,000 \times 0.412}{0.063}\right)^{0.8} \times \left(\frac{0.535 \times 0.063}{0.0248}\right)^{0.4}$$

$$= 0.0009 \times 2,490,000^{0.8} \times 1.36^{0.4}$$

$$= 0.0009 \times 130,900 \times 1.132$$

$$= 133 \ \text{Btu/(hr)(sq ft)(°F)}$$

7-13. Modifications of Nusselt's Expression. Numerous modifications of Nusselt's expression have been published by different experimenters—some with a view toward greater accuracy of results but in most cases for the purpose of simplifying or shortening calculations when the expression is applied to some specific fluid or to certain variations in the properties of that fluid.

Attention has already been called to the need for a correction factor when Nusselt's expression is applied to the calculation of the surface coefficient for a pipe having a small ratio of length to inside diameter and to allowance that must be made for rough and coated surfaces.

A number of the modifications frequently encountered are presented in the following articles.

7-14. Gases in Tubes. It has already been observed that the Prandtl number $c_p\mu/k$ for gases is most commonly of the order of 0.65 to 0.84. These numbers when raised to the 0.4 power, as applied in Eq. (7-5), are frequently considered to be sufficiently close to unity to justify their omission from Nusselt's equation when it applies to a gas. As a general practice this procedure is not to be recommended, for the difference between unity and such a value for $(c_p\mu/k)^{0.4}$ as $0.65^{0.4}$ is almost 16 per cent.

For calculations applying to the heating or cooling of a specific gas in a limited temperature range, such as air at temperatures of 0 to 2000°F, where $c_p\mu/k$ is close to 0.68 and $(c_p\mu/k)^{0.4}$ is close to 0.86, Eq. (7-5) may well be modified to such a form as

$$h_c = 0.023 \frac{k}{D} \left(\frac{DV\rho}{\mu}\right)^{0.8} \times 0.86$$

or

$$h_c = 0.020 \frac{k}{D} \left(\frac{DV\rho}{\mu}\right)^{0.8} \qquad \text{Btu/(hr)(sq ft)(°F)} \qquad (7\text{-}12)$$

For similar calculations dealing with a specific gas but in a more limited range of temperatures, Eq. (7-5) may be further simplified by substitution of appropriate average values of c_p, k, and μ into the equation, so that it becomes

$$h_c = \frac{C}{D} (DG)^{0.8}$$

or

$$h_c = C \frac{G^{0.8}}{D^{0.2}} \qquad \text{Btu/(hr)(sq ft)(°F)} \qquad (7\text{-}13)$$

where G = mass velocity (ρV), $lb_m/(sq ft)(hr)$, and C = coefficient, equal to $0.023k(1/\mu)^{0.8}(c_p\mu/k)^{0.4}$, or $0.023(k^{0.6}c_p{}^{0.4}/\mu^{0.4})$. Values for this coefficient C for *air* at various temperatures are shown in Table 7-2.

TABLE 7-2

Temp., °F.....	50	100	150	200	300	400	500
C............	0.00361	0.00368	0.00376	0.00383	0.00397	0.00411	0.00424

Temp., °F.....	600	800	1000	1200	1400	1600	1800
C............	0.00434	0.00460	0.00485	0.00501	0.00513	0.00531	0.00549

Values of $D^{0.2}$ for tubes and pipes of various sizes are shown in Table 7-3.

The application of Eq. (7-13) is shown in the following example.

Example 7-3.   Determine the surface coefficient of convection, in Btu/ (hr)(sq ft of inside pipe surface)(°F of temperature difference), when air at 100°F and 29.921 in. Hg pressure is heated while flowing through a 4-in. standard-weight steel pipe at a velocity of 1,000 fpm.

Solution.   By Eq. (7-13)

$$h_c = C \frac{G^{0.8}}{D^{0.2}} \qquad \text{Btu/(hr)(sq ft)(°F)}$$

$$C = 0.00368 \qquad\qquad \text{(Table 7-2)}$$
$$G = \rho V = 0.0708 \times 1000 \times 60 = 4,248$$
$$D^{0.2} = 0.804 \qquad\qquad \text{(Table 7-3)}$$
$$h_c = \frac{0.00368 \times 4248^{0.8}}{0.804} = 3.56 \text{ Btu/(hr)(sq ft)(°F)}$$

TABLE 7-3

Tubes, actual inside diameter, in.	$D^{0.2}$, ft	Standard-weight pipe		$D^{0.2}$, ft
		Nominal inside diameter, in.	Actual inside diameter, in.	
0.25	0.461	¼	0.364	0.496
0.375	0.500	⅜	0.493	0.528
0.50	0.530	½	0.622	0.553
0.75	0.574	¾	0.824	0.585
1.00	0.608	1	1.049	0.614
1.25	0.636	1¼	1.380	0.649
1.50	0.660	1½	1.610	0.669
2.00	0.699	2	2.067	0.703
2.50	0.731	2½	2.469	0.729
3.00	0.758	3	3.068	0.761
4.00	0.803	4	4.026	0.804
5.00	0.839	5	5.047	0.841
6.00	0.871	6	6.065	0.872
8.00	0.922	8	7.981	0.922

7-15. Limitations of Simplified Expressions. The practice of using simplified forms of Eq. (7-5), applied to a limited range of temperatures, needs caution. It may be observed, for example, that Eq. (7-13) is fundamentally the same as Eq. (7-5), which is applicable not only to gas but also to liquid films. Equation (7-13), however, should not be used for calculating the conductance of a liquid film, except at a definitely fixed temperature. A change in temperature changes the density, specific heat, thermal conductivity, and viscosity of a liquid in different measure from the change produced in a gas; and as already observed, an increase in temperature effects a decrease in the viscosity of a liquid but an increase in the viscosity of a gas.

The effect of a change in temperature upon the film conductance may be illustrated by applying Eq. (7-5) for calculating the surface coefficients of air and water, both at 60 and at 80°F. The change from 60 to 80° is found to increase the film conductance for air by less than 1 per cent, but it increases the conductance for water by more than 11 per cent.

7-16. Water in Tubes. The surface coefficient of convection for the heating or cooling of water in tubes at temperatures not exceeding 180°F may be expressed quite accurately by the simplified formula

$$h_c = 0.00134(t + 100) \frac{V^{0.8}}{D^{0.2}} \quad \text{Btu/(hr)(sq ft)(°F)} \quad (7\text{-}14)$$

where t = average water temperature, °F.

V = velocity, ft/hr.

D = inside diameter, ft.

If the temperature drop across the inside film is estimated to be more than 10°, t should be the mean temperature of the film.

7-17. Oil in Pipes. For the heating of oils in pipes, Morris and Whitman's[8] results may be approximated quite closely by means of the empirical formula

$$h_c = 0.034 \frac{V}{\mu^{0.63}} \quad \text{Btu/(hr)(sq ft)(°F)} \quad (7\text{-}15)$$

where V = velocity in feet per hour and μ = viscosity in pounds mass per foot-hour at the average temperature of the oil film. If along the length of the pipe the temperature limits are far apart, the viscosity should be the average viscosity of the temperature range, determined by integration.

If the oil is being cooled, the film conductance will be reduced about 25 per cent.

7-18. Liquids in Coiled Tubes. In the foregoing material, pipes have been assumed to be substantially straight. If tubes are helically coiled, turbulence will be increased and a consequent increase in film conductance is to be expected. An investigation by Richter[21] on water films in a double-pipe heat exchanger has shown the over-all heat transfer for the coiled tubes to be approximately 20 per cent greater than for straight tubes. Under the conditions of the tests the increase in film coefficient would be quite similar in percentage to the increase in over-all heat transfer.

7-19. Interdependence of Pressure Loss and Heat Transfer in Pipes and Tubes. In Art. 6-4, which dealt with the pressure loss in pipes and tubes, the relationships were noted not only between pressure loss and power requirements but also between pressure loss and heat transfer. Pressure loss and film conductance are so interdependent that substantially every change that may be made in the physical properties of the fluid or in the size, shape, or surface condition of the tube in an attempt to increase heat transfer results in an increase in pressure loss.   An exception to this usual relationship is that any coating or contamination of the tube wall decreases the rate of heat transfer although it generally increases the pressure loss.

In the design of heat-transfer apparatus the two factors ordinarily within the control of the designer are the diameter of the tube and the velocity of the fluid.   An increase in the velocity of the fluid effects an increase in film conductance in proportion to $V^{0.8}$.   An increase in the diameter of the tube reduces the film conductance in the ratio of $1/D^{0.2}$; and, for a fixed quantity of fluid flowing, an increase in the diameter means a decrease in the velocity and a further lowering of the film conductance. The use of small tubes and high velocities, therefore, is conducive to high rates of heat transfer, but not without limitation, since small tubes and high velocities call for increased power consumption in overcoming increased frictional resistance.   The designer, therefore, is confronted with the problem of effecting a suitable balance between the gain in heat transfer by the use of small tubes and the increase both in manufacturing cost and in operating expense.

7-20. Calculation of Film Conductance by Relationship to Pressure Loss.   When the relationship between film conductance and pressure loss is expressed in equation form it is possible to compute the film conductance by determination of the pressure loss.   This procedure is advantageous in cases where friction data can be obtained from experiments with less difficulty than heat-transfer data.   Conversely, calculations of heat transfer applied to the design of heat exchangers may be used in predictions of pressure loss.

Reynolds' analogy[22] between heat transfer and fluid friction in geometrically similar systems of piping (Art. 7-4), which was developed from the observation that heat and momentum in a fluid in laminar flow are transferred in the same manner, has produced a simple expression

$$N_{\text{st}} = \frac{f}{8}$$

but since the Stanton number

$$N_{\text{st}} = \frac{N_{\text{Nu}}}{N_{\text{Re}} \cdot N_{\text{Pr}}} = \frac{h_c}{V c_p \rho}$$

the film conductance (for certain limited applications) may be expressed as

$$h_c = \frac{f}{8} V c_p \rho \qquad \text{Btu/(hr)(sq ft)(°F)} \qquad (7\text{-}16)$$

where f = friction factor as shown in Fig. 6-4.

V = velocity, ft/hr.

c_p = specific heat at constant pressure.

ρ = density, lb_m/cu ft.

In deriving this expression, Reynolds assumed that the product of the specific heat and viscosity $c_p\mu$ is equal to the thermal conductivity k, or, in other words, the Prandtl number $c_p\mu/k$ equals unity—a relationship that is approximately true only for gases having more than three atoms per molecule.* For such gases Eq. (7-16) serves with fair accuracy, but for gases with a smaller number of atoms per molecule and a smaller value for the Prandtl number the equation is less accurate. For the flow of liquids, Eq. (7-16) is generally too inaccurate to be serviceable, the inaccuracy increasing with increase in the Prandtl number.

The dependence of film conductance upon the Prandtl number is readily observed by noting the slope of the graph in Fig. 7-1. Here, for $N_{\text{Pr}} = 1$, the ordinate $h_c/Vc_p\rho$, according to Eq. (7-16), is $f/8$. For $N_{\text{Pr}} = 1$, the three groups of curves for $N_{\text{Re}} = 10,000$, 25,000, and 50,000 show values of f which are substantially equal to the values of f shown in Fig. 6-4 for smooth pipe at these same Reynolds numbers. In contrast, for high values of N_{Pr}, the values of f are shown to be only a fraction of those found for $N_{\text{Pr}} = 1$.

Many researchers in the field of heat transfer have applied various extensions to Reynolds' analogy in attempts to make it applicable to turbulent instead of laminar flow of various fluids,

---

* The Prandtl number for gases increases with increase in the number of atoms per molecule, because of a decrease which occurs in the ratio of specific heats c_p/c_v. As shown in Art. 2-6, the thermal conductivity of a gas is related to its viscosity and to its specific heat at constant volume. This relationship is expressed by $k = a\mu c_v$, where a is a constant depending upon the number of atoms per molecule. Otherwise stated, $1/a = c_v\mu/k$. By multiplying both sides of the equation by the ratio c_p/c_v,

$$\frac{c_p}{ac_v} = \frac{c_p\mu}{k}$$

in tubes of various shapes and lengths. The results of these attempts are generally in the form

$$h_c = \frac{f}{8} V c_p \rho \psi \qquad (7\text{-}17)$$

where the function ψ accounts for the deviation of conditions from those upon which Reynolds' analogy was based. These expressions are useful but are subject to the limitation that ψ is in many cases a complicated expression and is generally based upon the assumption of highly turbulent flow in long tubes or ducts where the velocity and temperature profiles over the cross section do not vary along the length of the duct.

A simple expression for h_c in terms of the friction factor, which applies to turbulent flow of various fluids in long tubes or ducts, may be derived from the McAdams equation (7-5) as follows.

McAdams' equation may be written in the following forms:

$$h_c = 0.023 \frac{k}{D} N_{\mathrm{Re}}{}^{0.8} N_{\mathrm{Pr}}{}^{0.4}$$

$$= 0.023 \frac{k}{D} \frac{N_{\mathrm{Re}}}{N_{\mathrm{Re}}{}^{0.2}} \frac{N_{\mathrm{Pr}}}{N_{\mathrm{Pr}}{}^{0.6}}$$

$$= 0.023 \frac{k}{D} \frac{DV\rho}{\mu N_{\mathrm{Re}}{}^{0.2}} \frac{c_p \mu}{k N_{\mathrm{Pr}}{}^{0.6}}$$

$$= 0.023 \frac{V c_p \rho}{N_{\mathrm{Re}}{}^{0.2} N_{\mathrm{Pr}}{}^{0.6}}$$

But for smooth tubes with turbulent flow, f may be expressed[*] quite accurately as $0.184/N_{\mathrm{Re}}{}^{0.2}$, and therefore $0.023/N_{\mathrm{Re}}{}^{0.2}$ in the equation above may be replaced by $f/8$, resulting in the expression

$$h_c = \frac{f V c_p \rho}{8 N_{\mathrm{Pr}}{}^{0.6}} \qquad \text{Btu/(hr)(sq ft)(°F)} \qquad (7\text{-}18)$$

[*] The Fanning friction factor f_F is given by McAdams, "Heat Transmission," p. 155, McGraw-Hill Book Company, Inc., New York, 1954, equal to $0.046/N_{\mathrm{Re}}{}^{0.2}$, for the range of N_{Re}, 5,000 to 200,000, and by Stoever, "Applied Heat Transmission," p. 114, McGraw-Hill Book Company, Inc., New York, 1941, equal to $0.0653/N_{\mathrm{Re}}{}^{0.228}$, for the range of N_{Re}, 4,000 to 1,000,000. This factor, in both publications, is for substitution in the Fanning equation $\Delta p = 4 f_F (L/D)(\rho V^2/2g_c)$. Corresponding values for substitution in the Darcy-Weisbach equation $\Delta p = f(L/D)(\rho V^2/2g_c)$ are $0.184/N_{\mathrm{Re}}{}^{0.2}$ and $0.2612/N_{\mathrm{Re}}{}^{0.228}$.

The value of ψ, of Eq. (7-17), as developed from the McAdams equation, is seen to be $1/N_{\text{Pr}}^{0.6}$. Although Eq. (7-18) shows the relationship between the film conductance and the friction factor specifically for smooth tubes, it is commonly applied in calculations of the film conductance for flow through tubes of any degree of roughness, in keeping with an assumption that the film conductance varies in direct proportion to the friction factor—an assumption that is based upon inconclusive evidence.

If Eq. (7-18) had been developed from an equation such as Eq. (7-11) where the exponent of N_{Pr} is taken to be $\frac{1}{3}$, then ψ would be $1/N_{\text{Pr}}^{\frac{2}{3}}$. Use of this value of ψ by Colburn[23] and others is based not on any claim of better agreement with test results but merely on the observation that in expressions for film conductance various investigators have given the exponent of the Prandtl number a value ranging from 0.3 to 0.4, and $\frac{1}{3}$ represents a value within that range. Colburn's equation is

$$N_{\text{St}}N_{\text{Pr}}^{\frac{2}{3}} = \frac{h_c}{Vc_p\rho}\left(\frac{c_p\mu}{k}\right)^{\frac{2}{3}} = \frac{f_F}{2} = \frac{f}{8} \qquad (7\text{-}19)$$

where f_F is Fanning's friction factor and all other terms are as already defined.

The form of this equation is convenient for use in conjunction with graphs which show the relationships of the Stanton, Prandtl, and Reynolds numbers. It will be recalled that Fig. 7-1 shows plots of the Stanton versus the Prandtl number for three different values of the Reynolds number. Colburn has shown certain advantages of plotting $N_{\text{St}}N_{\text{Pr}}^{\frac{2}{3}}$ (which he calls j) versus N_{Re} for different values of the Prandtl number. The Stanton number N_{St} is a dimensionless parameter which combines the Nusselt, Reynolds, and Prandtl numbers in one term $N_{\text{Nu}}/N_{\text{Re}}N_{\text{Pr}}$, which equals $h_c/Vc_p\rho$. It has further significance in that it can be shown to represent test data in a form which requires no evaluation of the physical properties of the fluid. Heat transfer by convection is expressed as $h_c A \Delta t_m$, where h_c is the film conductance and Δt_m is the mean temperature difference across the film which coats the surface A of any given test length of tube. The heat transfer is also equal to $\rho A_1 Vc_p(t_1 - t_2)$, where $\rho A_1 V$ represents the mass of fluid flowing per unit of time through the tube of cross-sectional area A_1, and $c_p(t_1 - t_2)$ represents the heat added or removed in changing unit mass from an initial temperature

l_1 to a final temperature t_2. When these expressions for heat transfer are equated

$$h_c A \, \Delta t_m = \rho A_1 V c_p (t_1 - t_2)$$

and
$$\frac{h_c}{V c_p \rho} = \frac{(t_1 - t_2) A_1}{\Delta t_m A} \tag{7-20}$$

The right-hand side of this equation is seen to express the Stanton number in terms of test data.

Example 7-4. Determine, by use of the friction factor shown in Fig. 6-4, the surface coefficient in Btu/(hr)(sq ft of inside tube surface)(°F of temperature difference) for flow of water in a clean smooth horizontal tube of 2-in. diameter under the conditions stated in Ex. 7-1, for which the Reynolds number is 69,500 and the Prandtl number is 4.52.

Solution. For smooth-drawn tubing, the relative roughness ϵ/D, found from Table 6-1, is $0.000005 \times 12/2.00 = 0.00003$, and the friction factor f, found from Fig. 6-4, is 0.0194.

By Eq. (7-18)

$$\begin{aligned} h_c &= \frac{fV c_p \rho}{8 N_{Pr}^{0.6}} \\ &= \frac{0.0194 \times 11{,}100 \times 0.997 \times 61.99}{8 \times 4.52^{0.6}} \\ &= 673 \text{ Btu/(hr)(sq ft)(°F)} \end{aligned}$$

NOTE: This result is in close agreement with the answer to Ex. 7-1, solved by application of Eq. (7-5), namely, 686 Btu.

It may be observed that the amount of calculation involved in applying Eq. (7-18) is very similar to that required for Eq. (7-5). For solving either equation, Reynolds' and Prandtl's numbers must both be calculated. Equation (7-18), furthermore, requires measurements of pressure loss or the use of Fig. 6-4 for determining the friction factor, with the exception that for flow in smooth-drawn tubing the friction factor may be computed fairly accurately as $0.184/N_{Re}^{0.2}$. Equation (7-18) has the merit of emphasizing the effect of relative roughness of the pipe surface and of producing a result that is numerically dependent upon the surface roughness. As a further argument for the use of this equation it may be noted that the friction factor has been determined experimentally with greater precision than is generally possible in measurements of heat transfer. Those who prefer the use of such an expression as Eq. (7-5) may argue that little evidence has been produced to assure us that Reynolds' analogy may be applied with equal accuracy to flow in rough and in smooth pipes. Cope[24] has found in tests of air flow in tubes with three degrees of unusual roughness, artificially produced, that the friction increased by an amount several times the increase in surface conductance. He concluded that for the same pressure loss or power consumption, greater heat transfer was obtained with a smooth than with a rough tube. It may be further argued that since most forms of heat exchangers are made with smooth tubes for reasons beyond the matter of heat transfer, expressions such as Eq. (7-5), which involve no variations in friction factor, are generally applicable.

7-21. Heat Transfer and Pressure Loss in Heat Exchangers. If the surface conductance in heat exchangers is to be estimated from the pressure drop, the relationship between heat transfer and pressure drop should be known not only for the straight runs of tubes but for all other parts of the heat exchanger where heat is transferred. The resistances which a fluid may encounter in its path through a heat exchanger may include those which occur because of disturbances to the fluid stream at the inlet and outlet ends of tubes and those due to changes in velocity and direction of flow through curved tubes, return bends, and various shapes of construction. Attempts have been made and are still being made to extend the analogy between heat transfer and pressure drop beyond the type of flow that occurs in long tubes, but in some cases with questionable success. In some types of flow, such as where a fluid flows across a tube or a tube bank, no simple relationship between heat transfer and pressure drop is to be expected, for here forces are set up at the front and rear of the tube which are not related to heat transfer. Even for some types of flow for which a relationship between heat transfer and pressure drop may be expected, the determination of the equivalent diameter of the stream for use in establishing the Reynolds number and the friction factor are subject to considerable uncertainty. Generally, predictions of the heat-transfer performance of heat exchangers are more likely to be based directly upon calculations of heat transfer than upon pressure drop, even though a knowledge of pressure drop is essential for the determination of pumping heads and power consumption.

7-22. Forced Convection over Outside Surfaces. Heat exchange between a pipe and the fluid within it ordinarily takes place entirely by convection. Since there is no difference in the temperature of the pipe surface on opposite sides of the fluid stream, no radiation of heat from one portion of the surface to another can take place. Where heat is transferred through the film on the outside of pipes, however, radiation to or from surrounding surfaces may occur. The inside surfaces of the furnace of a brick-set water-tube boiler, for example, are usually more than 1000° above the temperature of the external surface of the water tubes. In this case more heat may be transferred by radiation than by convection; but in any event, the transfer by each of the two processes may be computed separately. In many types of heat exchangers, such as fan coils for heating or cooling air, the enclosing surfaces are at a temperature so nearly that of the coils that the amount of heat transferred by radiation is small and its calculation is usually neglected or included in the expression for film conductance, especially in cases where such expressions are derived from experimental work in which no attempt has been made to separate the amounts of heat transferred by convection and by radiation.

Investigations of the conductance of heat through films on the outside of pipes and other cylindrical surfaces have generally been limited to flow at right angles across pipes and wires, either singly or in banks. Most investigations have furthermore dealt only with the flow of air.

King and Knaus[25] express the surface coefficient for turbulent flow across a single pipe by the Nusselt equation in the form

$$h_c = 0.38 \frac{k}{D} N_{Re}^{0.56} N_{Pr}^{0.3} \qquad \text{Btu/(hr)(sq ft)(°F)} \qquad (7\text{-}21)$$

where D = outside diameter in feet and all other terms are the same as in Eq. (7-5).

Various experiments on flow across pipes have shown that the fluid film is not of uniform thickness, being thinner on the side facing the stream than on the downstream side. Because of this complication, equations of the Nusselt type have appeared to represent the surface coefficients in such cases less satisfactorily than empirical formulas developed from tests with specific fluids.

7-23. Flow of Gases across Wires and Tubes. An empirical formula recommended by King and Knaus[26]* for turbulent flow

---

* In this and in the following formula from the same source, consistent units have been substituted for the units originally employed.

of air across a single rod or tube is

$$h_c = 0.025 \frac{G^{0.58}(1 + 0.000576t)}{D^{0.42}} \qquad (7\text{-}22)$$

And for turbulent flow of air across tube banks, four or more rows deep, with staggered tube spacing,

$$h_c = 0.0032(t + 460)^{0.3} \frac{G^{2/3}}{D^{1/3}} \qquad (7\text{-}23)$$

where h_c = surface coefficient, Btu/(hr)(sq ft)(°F).

t = average air temperature, °F.

D = outside tube diameter, ft.

G = mass velocity through the minimum free area between the tubes, lb/(hr)(sq ft of free area).

Data that are more generally applicable for the cross flow of gases over tube banks, 10 rows of tubes in depth, have been developed by Grimison[27] for substitution into an expression for film conductance in the form

$$h_c = B \frac{k}{\mu^m} \frac{G^m}{D^{1-m}} \qquad (7\text{-}24)$$

where values of B and m, which depend upon the arrangement of tubes (in line or staggered) and their longitudinal and transverse spacing, are presented in tabular and in graphical form. Correction factors are to be applied for tube banks of a greater or lesser number of rows of tubes in depth.

Extensive data on the film conductance and frictional resistance for cross flow of gases over banks of plain and finned tubes of various shapes and arrangements have been presented by Kays and London[28] in the form of plots of $N_{St}N_{Pr}^{2/3}$ versus N_{Re}. The film conductance h_c may readily be found from these plots for any value of $N_{Pr}^{2/3}$ by multiplying the Stanton number N_{St} by the mass velocity G at the minimum area between the tubes and by the specific heat c_p. It should be noted that in applications of the value of h_c to calculations of the rate of heat transfer to or from the surface of finned tubes, account must be taken of the variations in temperature gradients along the fins which cause the effectiveness of the fin surface to be less than that of the tube surface. Compensation for the reduction in temperature differential is usually made by multiplying h_c by a factor termed the fin efficiency, or fin effectiveness. This factor, which is the ratio of the average temperature difference over the extended surface to that over the basic surface, may represent the effectiveness of an individual fin, or in the case of multiple fins it may be modified so as to represent a weighted average of the effectiveness of the fins and of the surface to which they are attached. Plots of fin effectiveness for fins of various depths and thicknesses are shown by Kays and London.[28]

7-24. Flow of Gases Parallel to Plane Surfaces. Experiments on the flow of air as a cooling medium, flowing parallel to smooth plane surfaces, indicate that the surface coefficient for turbulent flow may be expressed by the equation

$$h_c = 0.055 \frac{k}{L} \left(\frac{LV\rho}{\mu} \right)^{0.75} \tag{7-25}$$

where all units are consistent, as on the hour-foot-pound basis. This equation, although based upon tests with atmospheric air only and with the length of the surface L only about 1 ft, has been developed by dimensional analysis [in the Nusselt form, modified in the manner of Eq. (7-12)] and is accordingly considered applicable for calculations of heat transfer from smooth plane surfaces to any gas over a considerable range of variation in physical properties. Beyond a length of 2 ft it is probable that turbulence will be fairly constant, and the value of L to be substituted into the equation in the case of long surfaces should be limited to 2.

An investigation of the surface coefficients for surfaces of various building materials has been made by Rowley, Algren, and Blackshaw.[29] Their results are shown in Fig. A-10. Values read from the graph include not only the conductance due to convection but also the heat transferred by radiation. Although the tests were made at only one mean air temperature in the range of 65 to 70°F, the results are considered applicable to all ordinary problems encountered in the calculation of the heating and cooling requirements of buildings.

7-25. Forced Convection—Streamline Flow. In most of the forms of heat-exchange apparatus where convection is forced, i.e., made positive, the velocity of the fluid is above the critical velocity and the flow is accordingly turbulent. In a lesser number of instances, however, particularly in the case of liquids, the fluid is moved at definite low velocities that are less than the critical, where, if no heat were being added or removed, the flow would be in streamline form. When heat is being transferred, natural or free convection currents are set up and the flow is not strictly streamline, although it is usually so termed.

Heat transfer in streamline flow has been investigated less extensively than in the range of turbulent flow, but from available experimental results it is concluded that no satisfactory correlation has yet been established among the results of the various investigators or between experimental results and theoretical formulas developed on the basis of the known factors influencing the flow.

Tests on the heating of water, glycerol, and several kinds of oil in streamline flow have produced results which have been shown by W. J. King[30] to be approximated by the formula

$$h_c = 2.53 \frac{k}{D} \left(\frac{Wc_p}{kL} \right)^{1/3} \qquad (7\text{-}26)$$

within the range of values of Wc_p/kL from 8 to 3,500, where

h_c = surface coefficient of convection, Btu/(hr)(sq ft) (°F of arithmetical mean temperature difference between the fluid and the tube wall).

W = rate of fluid flow, lb_m/hr.

D and L = diameter and length* of the heated surface, ft.

c_p = specific heat, Btu/(lb_m)(°F).

k = thermal conductivity of the fluid, Btu/(hr)(sq ft) (°F/ft).

For flow in a tube or pipe, where W is equal to $(\pi/4)D^2V\rho$, Eq. (7-26) may be expressed in the form

$$h_c = 2.34 \left(\frac{V\rho c_p k^2}{DL} \right)^{1/3} \qquad (7\text{-}27)$$

where ρ = density, lb_m/cu ft, and all other terms are the same as in the previous equation.

Limited data indicate a reduction of 30 per cent in the surface coefficient when oil is being cooled, this reduction being explained by the notable increase in viscosity and consequent decrease in convection that occurs when the temperature of the main body of the oil is lowered.

---

* For a discussion of the effect of tube length on film conductance for laminar flow, see Ref. 31.

For streamline flow of water in vertical pipes at velocities less than 0.1 ft/sec, Colburn and Hougen[32] derived the following empirical formulas for the water-film coefficient:

For the heating of water in upward flow:

$$h_c = 0.37 \; \Delta t^{\frac{1}{3}} t \qquad (7\text{-}28)$$

and for the heating of water in downward flow:

$$h_c = 0.44 \; \Delta t^{\frac{1}{3}} t \qquad (7\text{-}29)$$

where h_c = surface coefficient of convection, Btu/(hr)(sq ft)(°F).

Δt = temperature change across the water film, °F.

t = mean temperature of the film, °F.

The higher value of h_c by Eq. (7-29) than by Eq. (7-28) is due apparently to greater disturbance of the water film when the downward flow of the main body of the stream is opposed by the rising current that is induced in the film by the difference in densities of the main body and the warmer film.

Problems

1. A ½-in. pipe (actual i.d. = 0.62 in.) carries water at the rate of 100 lb/hr. If the water temperature is 200°F, is the flow streamline or turbulent?

2. Water flows in a 3-in. i.d. tube 5 ft long at a Reynolds number of 8,000. The Prandtl number is 5.1, and the k value equals 0.36. Calculate the coefficient of forced convection heat transfer.

3. Water at an average temperature of 180°F is heated while flowing at a velocity of 4,000 fph through a 2-in. standard-weight steel pipe. Determine the inside-surface coefficient of convection in Btu/(hr)(sq ft)(°F).

4. Check the result found for Prob. 3, applying Eq. (7-14).

5. Check the result found for Prob. 3, applying Eq. (7-8).

6. Determine the critical velocity for water flowing under the conditions of temperature and pipe size stated in Prob. 3.

7. Air at an average stream temperature of 700°F and at 14.7 psia is cooled while flowing at a velocity of 50 fps through a 4-in. i.d. clean horizontal tube 10 ft long. The average inside-wall temperature is 300°F. Determine the inside-surface coefficient of convection in Btu/(hr)(sq ft)(°F).

8. Water at an average stream temperature of 200°F is heated while flowing through a 3-in. i.d. clean horizontal tube at a velocity of 0.5 fps. The inside-surface temperature of the tube is 240°F. Determine the inside-surface coefficient of convection in Btu/(hr)(sq ft)(°F).

9. Water is heated while flowing through a 1-in. i.d. tube at a velocity of 1 fps. The average stream temperature of the water is 100°F and the average surface temperature of the tube wall is assumed to be 140°F. Determine (a) the inside-surface coefficient of convection and (b) the length of tube required for raising the temperature of the water 50°F as it passes through the tube.

10. A device used for heating water is made up of 3-in. i.d. tubes. The water flows through these tubes at a velocity of 1 fps. It is estimated that an inside-tube surface temperature of 200°F will exist when the mean stream temperature of the water is 180°F. Determine the heating capacity in Btu/hr for 100 sq ft of tube surface.

11. Sulfur dioxide gas at an average temperature of 40°F and a density of 0.18 lb_m/cu ft is heated while flowing through a 3-in. i.d. tube at a velocity of 1,000 fpm. Determine the inside-surface coefficient of convection in Btu/(hr)(sq ft)(°F).

12. What would be the effect on film conductance for fluid flow within a pipe if the inside surface were made smoother while all other conditions remained constant?

13. Water at an average temperature of 160°F is cooled slightly while flowing at a velocity of 8,000 fph through a 10-ft length of clean horizontal smooth-drawn 1.50-in. i.d. copper tube. Determine (a) the friction factor, (b) the pressure drop in inches of water, and (c) the inside-surface coefficient of convection in Btu/(hr)(sq ft)(°F), by application of Eq. (7-18).

14. Water at an average temperature of 80°F is heated while flowing through a 20-ft length of 2-in. i.d. tubing maintained at a wall temperature of 100°F. If, for a velocity of 4 fps, the pressure drop in a 10-ft portion of the tube is found by test to be 3.75 in. of water, determine (a) the friction factor and (b) the inside-surface coefficient of convection in Btu/(hr) (sq ft)(°F).

15. Determine the length of 3-in. i.d. steel tube required to raise the temperature of 700 lb of water per hour 20°F if the average water temperature is 200°F and the average temperature of the inside surface of the tube is 400°F.

16. 200 lb of 100 per cent methyl alcohol (methanol) flows per minute at an average temperature of 90°F through a long 4- by 2-in. steel conduit which is maintained at an average temperature of 96°F. The specific gravity of the fluid is 0.80, and the specific heat is 0.58. Determine the inside-surface coefficient of convection by (a) the McAdams equation and by (b) a suitable expression involving the friction factor.

17. Air at an average temperature of 70°F and at 14.7 psia flows across a 1-in. pipe (o.d. = 1.315 in.) at a velocity of 1,000 fpm. Determine the outside-surface coefficient of convection by application of Eq. (7-21).

18. Check the result found for Prob. 17, applying Eq. (7-22).

19. Air at atmospheric pressure and an average stream temperature of 160°F is heated while flowing through a 6-in. i.d. clean horizontal tube 10 ft long at a velocity of 1,000 fpm. The heating medium on the outside of the tube is atmospheric steam. Determine the inside-surface coefficient of convection (a) by application of Eq. (7-13) and (b) by application of Eq. (7-18).

20. Water at an average stream temperature of 220°F is heated while flowing through a 3-in. clean horizontal tube 5 ft long at a velocity of 0.02 fps. The temperature of the inside surface of the tube is 260°F. Determine the inside-surface coefficient of convection in Btu/(hr)(sq ft)(°F).

21. 700 lb of water is heated per hour as it flows through a 6- by 10-in. conduit 25 ft long. If the temperatures of the water at the two ends of the conduit are 158 and 162°F, what must be the average temperature of the inside surface of the conduit?

22. Water at an average temperature of 100°F is heated while flowing at a velocity of 200 fph through a 2-in. standard-weight steel pipe 10 ft long. Determine the inside-surface coefficient of convection in Btu/(hr) (sq ft)(°F).

23. One thousand cubic feet of air at 14.7 psia is heated per minute while flowing at an average temperature of 100°F through a 6- by 12-in. rectangular metal duct. The inside-surface temperature of the duct is 300°F. Determine the inside-surface coefficient of convection in Btu/(hr)(sq ft) (°F).

REFERENCES

1. McAdams, W. H.: "Heat Transmission," 3d ed., p. 152, McGraw-Hill Book Company, Inc., New York, 1954.
2. Conditions at the Surface of a Hot Body Exposed to the Wind, *Natl. Advisory Comm. Aeronaut. Tech. Rept.* 2; *Tech. Mem.* 272, p. 423, May, 1916.
3. Bemerkung über den Wärmeübergang im Rohr, *Physik. Z.*, **29**, 487 (1928).
4. The Analogy between Fluid Friction and Heat Transfer, *Trans. ASME*, **61**, 705–710 (1939).
5. Discussion of Paper by Th. von Kármán by Boris A. Bakhmeteff, *Trans. ASME*, **62**, 551–553 (1940).
6. Boelter, L. M. K., R. C. Martinelli, and H. Jonassen: *Trans. ASME*, **63**, 447–455 (1941).
7. *Z. Ver. deut. Ing.*, **53**, 1750 (1909).
8. Morris, F. H., and W. G. Whitman: Heat Transfer for Oils and Water in Pipes, *Ind. Eng. Chem.*, **20**, 234 (1928).
9. McAdams, W. H.: "Heat Transmission," 3d ed., p. 219, McGraw-Hill Book Company, Inc., New York, 1954.
10. The Analogy between Fluid Friction and Heat Transfer, *Trans. ASME*, **61**, 705–710 (1939).
11. On the Coefficient of Heat Transfer from the Internal Surface of Tube Walls, *Proc. Roy. Soc. London*, A, **127**, 540 (1930).
12. Martinelli, R. C.: Heat Transfer to Molten Metals, *Trans. ASME*, **69**, 947–959 (1947).
13. Nikuradse, J.: *Proc. Intern. Congr. Appl. Mech.*, 3d Congr., Stockholm, 1930, **1**, 239 (1931).
14. Lyon, R. N.: Liquid Metal Heat-transfer Coefficients, *Chem. Eng. Prog.*, **47**, 75–79 (1951).
15. Sieder, E. N., and G. E. Tate: *Ind. Eng. Chem.*, **28**, 1429 (1936).
16. Buxton, O. E., Jr.: Master's thesis in Mechanical Engineering, The Ohio State University, Columbus, Ohio, 1951.
17. *Bur. Standards Tech. Paper* 211, p. 326, 1922.

18. Kratz, A. P., H. J. Macintire, and R. E. Gould: *Univ. Ill. Bull.* 209 (1930).
19. Luke, G. E.: The Cooling of Electric Machines, *Trans. AIEE,* **42,** 636 (1923).
20. Moore, A. D.: "Heat Transfer Notes for Electrical Engineering," p. 74, George Wahr, Ann Arbor, Mich., 1938.
21. Richter, G. A.: Double Pipe Heat Interchangers, *Amer. Inst. Chem. Eng.,* **49,** pt. II, p. 147 (1919).
22. On the Extent and Action of the Heating Surface for Steam Boilers, *Proc. Manchester Lit. Phil. Soc.,* **14,** 7 (1874), reprinted in "Heat Transmission by Radiation, Conduction and Convection" by R. Royds.
23. Colburn, A. P.: *Trans. Am. Inst. Chem. Engrs.,* **29,** 174–210 (1933).
24. Cope, W. F.: *Proc. Inst. Mech. Engrs. (London),* **145,** 99–105 (1941).
25. "Refrigerating Data Book," pp. 144–145, American Society of Refrigerating Engineers, New York, 1942.
26. "Refrigerating Data Book," p. 89, American Society of Refrigerating Engineers, New York, 1937–1938.
27. Grimison, E. D.: Correlation and Utilization of New Data on Flow Resistance and Heat Transfer for Cross Flow of Gases over Tube Banks, *Trans. ASME,* **59,** 583–594 (1937).
28. Kays, W. M., and A. L. London: "Compact Heat Exchangers," National Press, Palo Alto, Calif., 1955.
29. Rowley, F. B., A. B. Algren, and J. L. Blackshaw: Surface Conductance as Affected by Air Velocity, Temperature and Character of Surface, *Trans. ASHVE,* vol. 36 (1930).
30. King, W. J.: *Mech. Eng.,* **54,** 412 (June, 1932).
31. Eckert, E. R. G.: "Introduction to the Transfer of Heat and Mass," pp. 98–105, McGraw-Hill Book Company, Inc., New York, 1950.
32. Colburn, A. P., and O. A. Hougen: *Ind. Eng. Chem.,* **22,** 522 (1930).

CHAPTER 8

FREE OR NATURAL CONVECTION

8-1. General. The contact of any fluid with a hotter surface reduces the density of the fluid and causes it to rise. If the surface is colder than the fluid, the fluid falls. Circulation thus set up by the difference in the temperature of a fluid and the surface with which it is in contact is called *free*, or *natural*, convection. If the fluid is a liquid, practically all the heat removed from a hotter surface is carried away by convection currents; but if it is a gas, the heat is removed both by convection and by radiation to surrounding surfaces. Since the two processes are dependent upon different factors, they should be computed separately.

The flow of free convection currents of air over a hot vertical plane surface has been investigated by a number of observers who have measured the velocities and temperatures at various heights and distances from the surface. They have found that the highest temperatures occur toward the top of and close to the surface. The highest velocity is found at a distance of about $\frac{1}{16}$ to $\frac{1}{8}$ in. from the surface. Air is drawn into the moving stream in a substantially horizontal direction at the bottom of the plate and more nearly diagonally upward at the top. The upward velocity increases along the height of the plate, but for tall plates no appreciable increase in either velocity or temperature is observed above a height of about 2 ft.

The difference in temperature between the upper portion of the plate and the air that has been warmed during its passage upward is obviously less than at the bottom, and the surface conductance in general decreases along the height of the plate. At a height of some 12 to 18 in., however, the effect of increased velocity apparently more than offsets the decreased temperature differential and the surface conductance increases slightly.

161

Above a height of about 2 ft the surface conductance remains practically constant.

8-2. Equations for Film Conductance.   In the development of an expression for the film conductance for natural convection it may be observed that the same factors are involved as in Nusselt's equation [Eq. (7-3)], which applies to forced convection, except that the velocity V is in turn dependent upon the force due to the buoyancy of the fluid, a force produced by the difference in temperature of the surface and the main body of the fluid and by the thermal expansion of the fluid.

If the difference in temperatures of the surface and the fluid is Δt, the average temperature of the fluid near the surface is $(\Delta t/2)°$ warmer or colder than the surrounding fluid.   If β is the coefficient of thermal expansion of the fluid, the change in volume $1/\rho$ for a temperature change of $(\Delta t/2)°$ is $(1/\rho)\beta(\Delta t/2)$, or the change in density is equal to $\rho\beta(\Delta t/2)$.   This value, in pounds mass per cubic foot, when multiplied by g/g_c, represents the buoyant force per cubic foot.   The work done per cubic foot when this force acts through a height of surface L equals $(g/g_c)\rho\beta(\Delta t/2)L$; and if all this work is converted into kinetic energy of motion in producing a velocity V,

$$\frac{g\rho\beta}{g_c}\frac{\Delta t}{2}L = \frac{\rho V^2}{2g_c} \tag{8-1}$$

from which

$$V^2 = g\beta\,\Delta t\,L \tag{8-2}$$

In order that this value of V^2 may be substituted into the Nusselt equation [Eq. (7-3)], that equation may be more conveniently expressed in the form

$$\frac{h_c L}{k} = C\left(\frac{L^2 V^2 \rho^2}{\mu^2}\right)^{b/2}\left(\frac{c_p\mu}{k}\right)^d \tag{8-3}$$

where the dimension L refers to the vertical height of plane surfaces or the diameter of horizontal pipes or wires, and the Reynolds number has been written as $(L^2 V^2 \rho^2/\mu^2)^{1/2}$ in order to present the velocity term to the second power.

Now, by substitution of the value of V^2 from Eq. (8-2) into Eq. (8-3), the expression becomes

$$\frac{h_c L}{k} = C\left(\frac{g\beta\,\Delta t\,L^3\rho^2}{\mu^2}\right)^{b/2}\left(\frac{c_p\mu}{k}\right)^d \tag{8-4}$$

The term $g\beta \, \Delta t \, L^3\rho^2/\mu^2$ is known variously as the Grashof number, Grashof's modulus, or the free-convection modulus. h_cL/k and $c_p\mu/k$ have already been defined as the Nusselt and the Prandtl numbers, or moduli, respectively. All three moduli are dimensionless.

The results of tests of free convection with various fluids, both liquid and gas, flowing over single horizontal pipes and wires and over vertical planes, show the exponents $b/2$ and d to be numerically the same. Since this is so, Eq. (8-4) may be written in the form

$$\frac{h_cL}{k} = C \left(\frac{g\beta \, \Delta t \, L^3\rho^2c_p}{\mu k} \right)^d \tag{8-5}$$

or

$$h_c = C \frac{k}{L} \left(\frac{g\beta \, \Delta t \, L^3\rho^2c_p}{\mu k} \right)^d \qquad \text{Btu/(hr)(sq ft)(°F)} \tag{8-6}$$

For convenience, the product of the Grashof and the Prandtl numbers is often expressed as $aL^3 \, \Delta t$, where a has been substituted for the fraction $g\beta\rho^2c_p/\mu k$. With this substitution, Eq. (8-6) may be written in the form

$$h_c = C \frac{k}{L} (aL^3 \, \Delta t)^d \qquad \text{Btu/(hr)(sq ft)(°F)} \tag{8-7}$$

The values to be substituted into Eq. (8-6) in applying it to any problem are to be evaluated at the mean temperature of the surface and the main body of the fluid. The coefficient of thermal expansion β for *gases* is usually considered, in keeping with the law of perfect gases, to be the reciprocal of the absolute temperature. All terms must be in consistent units, such as Btu, degrees Fahrenheit, pounds, feet, and hours. The value of g in feet per hour per hour is 4.17×10^8. Δt is the difference in temperature between the surface and the main body of the fluid. L has already been defined as the height of plane surfaces or the diameter of horizontal pipes or wires. It is quite apparent that for free convection over a long vertical plane or a horizontal cylinder the vertical dimension is the significant one. In the case of small short cylinders, either horizontal or vertical, and short vertical planes and spheres, free convection is influenced not only by the vertical but also by the horizontal dimension, and here experimental results have shown that the significant dimension L can be found satisfactorily by the expression

$$\frac{1}{L} = \frac{1}{L_{\text{horiz}}} + \frac{1}{L_{\text{vert}}} \tag{8-8}$$

For a sphere, therefore,

$$\frac{1}{L} = \frac{1}{D} + \frac{1}{D} = \frac{2}{D}$$

or L equals the *radius* of the sphere.

The exponent d of Eqs. (8-5) and (8-6) has been determined by a number of investigators by observing the slope of the graph that results when the Nusselt number, as found from test data, is plotted on logarithmic coordinate paper against the product of the Grashof and the Prandtl numbers, $N_{\text{Gr}} \times N_{\text{Pr}}$, which in Eqs. (8-5) and (8-6) is expressed as $g\beta \, \Delta t \, L^3 \rho^2 c_p / \mu k$. Figure 8-1

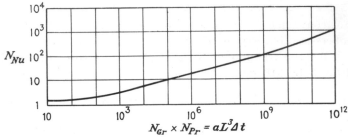

FIG. 8-1. Free convection for air over vertical plates.

shows such a graph, plotted from data of Saunders.[1] In the range most commonly encountered in practice, where $N_{\text{Gr}} \times N_{\text{Pr}}$ is between approximately 1,000 and 10^9, the line is almost straight and at a slope of $\frac{1}{4}$. At values of $N_{\text{Gr}} \times N_{\text{Pr}}$ higher than 10^9, encountered mainly in convection of liquids, the slope of the graph indicates a value of d more nearly equal to $\frac{1}{3}$. At values of $N_{\text{Gr}} \times N_{\text{Pr}}$ less than 1,000 (more commonly applying to convection over very fine wires with small temperature differences), d diminishes with decrease in $N_{\text{Gr}} \times N_{\text{Pr}}$, approaching zero when $N_{\text{Gr}} \times N_{\text{Pr}}$ is $\frac{1}{10,000}$ or smaller.

The foregoing values of d, recommended for use in Eqs. (8-5) and (8-6), are shown in tabular form in Table (8-1).

Values of the modulus a are shown for water at various temperatures in Table A-1 and Fig. A-9 and for air in Table A-2 and Fig. A-2. Figure A-9 shows values also for ethyl alcohol and for transformer oil.

The constant C of Eqs. (8-5) and (8-6) has been established by the work of a goodly number of investigators in tests of free convection over horizontal pipes and wires and over long vertical cylinders and vertical plates, cooled both by gases and by liquids. In the range where $N_{Gr} \times N_{Pr}$ is between 1,000 and 10^9, C has been given a general value of 0.55. The correlation of the results found by the different observers is not close enough to warrant drawing distinctions with great assurance between the values of C for pipes, wires, and plates in various positions; but from a number of tests, mainly limited to the flow of air over surfaces in different positions, it has been observed that the film conductance is slightly higher for vertical plates and for long

TABLE 8-1

$\dfrac{g\beta\,\Delta t\,L^3\rho^2 c_p}{\mu k} = (aL^3\,\Delta t)$	d
$< 10^3$	$< \frac{1}{4}$
$10^3 - 10^9$	$\frac{1}{4}$
$> 10^9$	$\frac{1}{3}$

vertical cylinders than for horizontal cylinders. Horizontal plates with the warm side facing upward transfer about twice as much heat per degree of temperature difference as do similar plates with the warm side facing downward.

Table 8-2 shows values of C to be applied to the calculation of film conductances for surfaces of various shapes and in different positions, in preference to the general value of 0.55.

Where $aL^3\,\Delta t$ lies between 10^3 and 10^9, Eq. (8-7) becomes

$$h_c = C\,\frac{k}{L}\,(aL^3\,\Delta t)^{\frac{1}{4}} \qquad \text{Btu/(hr)(sq ft)(°F)} \qquad (8\text{-}9)$$

The value of $aL^3\,\Delta t$ is likely to be more than 10^9 if the height L of the surface in air or other gases is 2 ft or more, or if the height of the surface in a liquid is more than 0.15 ft. For these cases W. J. King[2] recommends use of the formula

$$h_c = C_1\,\frac{k}{L}\,(aL^3\,\Delta t)^{\frac{1}{3}} \qquad (8\text{-}10)$$

or, since $(L^3)^{\frac{1}{3}}/L = 1$, this equation may be written also as

$$h_c = C_1 k(a\,\Delta t)^{\frac{1}{3}} \qquad \text{Btu/(hr)(sq ft)(°F)} \qquad (8\text{-}11)$$

where C_1 is assigned the general value of 0.13. Modifications of this value to be applied to various shapes and positions of the surface are found in Table 8-2.

Any dimension, either vertical or horizontal, greater than about 2 ft, relating to a surface surrounded by a gas, has a negligible effect upon the film conductance; and in the case of a horizontal pipe the effect of increasing the diameter beyond 6 or 8 in. is small. This observation has been made in experimental studies of free convection of gases but appears to be applicable also to free convection of liquids. The film conductance, calculated by Eq. (8-11), may be observed to be independent of the dimension

TABLE 8-2

$aL^3 \, \Delta t$, or $\dfrac{g\beta \, \Delta t \, L^3 \rho^2 c_p}{\mu k}$ (where L is limited to 2 ft)	$10^3\text{--}10^9$	$> 10^9$
Shape and position	C	C_1
Vertical plates..............................	0.55	0.13
Horizontal cylinders (pipes and wires)........	0.45	0.11
Long vertical cylinders......................	0.45–0.55*	0.11–0.13*
Horizontal plates, warm side facing upward...	0.71	0.17
Horizontal plates, warm side facing downward.	0.35	0.08
Spheres (L = radius).......................	0.63	0.15

* Turbulence and fluctuating eddies in the flow over long vertical cylinders cause uncertainty in the values of C and C_1.

L. Similarly, if, in Eq. (8-9), L is $\frac{1}{2}$ ft or more, the value of $(L^3)^{1/4}/L$ is not far from 1 and any increase in the dimension will have little effect upon the calculated value of the film conductance. It may be concluded that the value of L to be applied in any of the expressions for the film conductance due to free convection should not exceed 2 ft.

The values of C and C_1 for horizontal plates, shown in Table 8-2, apply when L represents the height of the plate if placed in a vertical position; but for plates of height exceeding 2 ft, L must be limited to 2.

Where repeated calculations of the surface coefficient h_c are made by application of Eq. (8-6) the procedure is obviously tedious, but for specific fluids the work may be greatly reduced if the product of the terms that represent the physical properties at the existing temperatures is read from a graph or tabulation of the modulus a.   Values of a for air at atmospheric pressure and at different temperatures are shown in Table A-2 and Fig. A-2. For other air pressures, a must be multiplied by the square of the pressure in atmospheres.   Values of k, which is independent of the pressure, also are shown in Table A-2 and Fig. A-2.

Figure A-9 shows values of a and k for water, ethyl alcohol, and transformer oil.

The following examples illustrate the application of Eqs. (8-9) and (8-11).

Example 8-1.   A 6-in. horizontal wrought-iron pipe, supplied with saturated steam at 250°F, is exposed to air at 70°F.   Determine the outside-surface coefficient of free convection in Btu/(hr)(sq ft of external surface) (°F of temperature difference).

Solution.   The mean temperature of the pipe surface and the main body of air will be approximately $(250 + 70)/2 = 160°F$ (neglecting the small difference in the temperatures of the steam and the outer surface of the pipe).

$$\text{At } 160°F, \, a_{\text{air}} = 0.77 \times 10^6, \text{ or } 770{,}000 \qquad \text{(Table A-2)}$$
$$L = 6.625 \text{ in.}/12 = 0.552 \text{ ft}$$
$$\Delta t = 250 - 70 = 180°F$$
$$aL^3 \, \Delta t = 770{,}000 \times 0.552^3 \times 180 = 23{,}200{,}000$$

Since $aL^3 \, \Delta t$ is less than 10^9, Eq. (8-9) is applicable.

$$C \text{ (by Table 8-2)} = 0.45$$
$$k \text{ for air at } 160°F \text{ (by Table A-2)} = 0.0172$$

$$h_c = C \frac{k}{L} (aL^3 \, \Delta t)^{1/4}$$

$$= 0.45 \frac{0.0172}{0.552} (23{,}200{,}000)^{1/4}$$

$$= 0.45 \times 0.0311 \times 69.4 = 0.97 \text{ Btu/(hr)(sq ft)(°F)}$$

NOTE: Published data on the over-all transfer coefficient for a 6-in. bar-pipe, under the conditions stated in this example, show a value of 2.5 Btu/(hr)(sq ft)(°F).   The heat given off by radiation, therefore, apparently is more than that transferred by convection.

Example 8-2.   A 3-in. horizontal wrought-iron pipe, supplied with saturated steam at 210°F, is immersed in water at 150°F.   Determine the surface coefficient of free convection, in Btu/(hr)(sq ft of external surface) (°F).

Solution.   The mean temperature of the surface and the main body of water will be approximately $(210 + 150)/2 = 180°F$ (neglecting the small difference in temperature of the steam and the outer surface of the pipe).

$$\text{At } 180°F, \, a_{\text{water}} = 17.1 \times 10^8 \qquad \text{(Table A-1)}$$
$$L = 3.500 \text{ in.}/12 = 0.292 \text{ ft}$$
$$\Delta t = 210 - 150 = 60°F$$
$$aL^3 \, \Delta t = 17.1 \times 10^8 \times 0.292^3 \times 60 = 2.53 \times 10^9$$

Since $aL^3 \Delta t$ is greater than 10^9, Eq. (8-11) is applicable.

k for water at 180°F (by Table A-1) = 0.389

$$h_c = 0.11k(a \, \Delta t)^{\frac{1}{3}}$$
$$= 0.11 \times 0.389(17.1 \times 10^8 \times 60)^{\frac{1}{3}}$$
$$= 0.11 \times 0.389 \times 4{,}680 = 200 \text{ Btu/(hr)(sq ft)(°F)}$$

SPECIFIC FREE CONVECTION

8-3. Vertical Plane Surfaces in Air. When problems in film conductance due to free convection relate specifically to the flow of air at atmospheric pressure, in a definite range of temperature, over vertical plane surfaces less than 2 ft in height, they may be solved by Eq. (8-9) further simplified by the substitution of a suitable value for $Cka^{\frac{1}{4}}$ into the expression.

At a mean temperature of 120°F, considered as the arithmetical mean of the surface and ambient air temperature, $ka^{\frac{1}{4}} = 0.524$ and for $C = 0.55$ the equation becomes

$$h_c = 0.29 \left(\frac{\Delta t}{L}\right)^{\frac{1}{4}} \qquad \text{Btu/(hr)(sq ft)(°F)} \qquad (8\text{-}12)$$

The value of $ka^{\frac{1}{4}}$ in the range of usual air temperatures is fairly constant, ranging from 0.549 at 0°F to 0.489 at 250°F. For pressures other than atmospheric, since a varies as the square of the pressure in atmospheres and in Eq. (8-9) is raised to the $\frac{1}{4}$ power, it follows that h_c varies as the square root of the pressure in atmospheres.

For large vertical plane surfaces (over 2 ft high), the film conductance is somewhat uncertain because of turbulence, but an average value may be found by Eq. (8-11) expressed in simplest form by the substitution of an appropriate value for $C_1ka^{\frac{1}{3}}$. Thus for a mean temperature of 120°F

$$h_c = 0.22 \, \Delta t^{\frac{1}{3}} \qquad \text{Btu/(hr)(sq ft)(°F)} \qquad (8\text{-}13)$$

This expression, however, is limited in its application to a smaller range in temperature than is Eq. (8-12), because of greater variation of $a^{\frac{1}{3}}$ than of $a^{\frac{1}{4}}$ with change in temperature. For a mean temperature of 0°F Eq. (8-13) becomes

$$h_c = 0.25 \, \Delta t^{\frac{1}{3}} \qquad \text{Btu/(hr)(sq ft)(°F)} \qquad (8\text{-}14)$$

and for a mean temperature of 250°F

$$h_c = 0.19 \, \Delta t^{\frac{1}{3}} \qquad \text{Btu/(hr)(sq ft)(°F)} \qquad (8\text{-}15)$$

8-4. Cylindrical Surfaces in Air. The film conductance due to free convection of air at atmospheric pressure over horizontal wires and pipes ranging up to 1 ft in diameter is less than for vertical plane surfaces of similar height, in proportion to the relative values of the constant C shown in Table 8-2. Substitution of the appropriate value for C of 0.45 in the expression results in the formula, for a mean temperature of 120°F,

$$h_c = 0.24 \left(\frac{\Delta t}{L}\right)^{\frac{1}{4}} \quad \text{Btu/(hr)(sq ft)(°F)} \quad (8\text{-}16)$$

For a mean temperature of 0°F the formula is

$$h_c = 0.25 \left(\frac{\Delta t}{L}\right)^{\frac{1}{4}} \quad (8\text{-}17)$$

and for 250°F

$$h_c = 0.22 \left(\frac{\Delta t}{L}\right)^{\frac{1}{4}} \quad (8\text{-}18)$$

For air pressures other than atmospheric, h_c varies as the square root of the pressure in atmospheres.

Although some evidence exists that the film conductance for a vertical pipe is greater than for the same pipe in a horizontal position, the difference is usually considered to be small, and Eqs. (8-16), (8-17), and (8-18), in which L is the diameter in feet, are usually applied equally to vertical and to horizontal pipes.

8-5. Effect of Surroundings. When any heat-transfer surface is close to a wall or ceiling or any enclosing structure, some effect upon free convection may be expected. The magnitude of the effect upon film conductance has been studied by Boelter and Cherry[3] in tests in which plane surfaces simulating a wall, a ceiling, a combination of the two, and an enclosing tunnel were located at various distances from a cylinder of 2.33-in. diameter, chromium-plated so as to reduce the surface emissivity. In these tests the Schlieren method was employed, in which the rate of heat transfer by convection is measured by the deflection of a beam of light directed so as to be initially parallel to the cylindrical surface. Observations were made at ambient air temperatures in the range from 293 to 394°F only, and the results accordingly may not be strictly applicable to other temperatures; but in general they indicate no measurable effect of any surface located one cylinder diameter or more from the nearest element of the cylindrical surface. More specifically, the presence of a vertical wall exerted no effect until it was brought within one-fourth of a cylinder diameter from the cylinder, and at closer positions the effect was small. When the cylinder was surrounded by four identical plane surfaces in the form of a tunnel, as the size of the tunnel was decreased, the reduction in film conductance started when the plane surfaces were brought within one cylinder diameter of the cylindrical surface. When the distance was reduced to one-tenth of the cylinder diameter, the reduction in film conductance amounted to approximately 30 per cent. A ceiling alone and a ceiling combined with a side wall showed effects intermediate between those of the vertical wall and of the tunnel.

8-6. Vertical Air Spaces. A number of investigations of heat transfer between vertical surfaces separated by an air space have shown that if the width of the space is greater than about 1½ in., convection is independent of the width. If the width is reduced, convection falls off rapidly until, for a width of ⅛ in., air currents are so restricted that convection is practically eliminated (Fig. A-11). The total heat transfer across a ⅛-in. air space may be considerably more than across a 1½-in. space, but the increase is due to increased conduction through the thinner barrier of relatively stagnant air and not due to convection. Radiation from one surface to the other takes place independently of the width of the space, provided the surfaces are large in comparison with the distance between them.

When an air space is used for heat insulation, it is most effective when it is divided into a number of narrow spaces by several separate layers of foil or thin sheets of bright metal having low radiating capacity. The advantage of this construction lies in the reduction of heat transfer both by convection and by radiation.

Problems

1. Air at atmospheric pressure and at a temperature of 60°F flows by free convection over the surface of a 3-in. horizontal steel pipe (o.d. = 3.500 in.) maintained at a surface temperature of 180°F. Determine the outside-surface coefficient of free convection in Btu/(hr)(sq ft)(°F).

2. Repeat Prob. 1, but for a vertical plate 3 ft high.

3. Check the result found for Prob. 2, applying from the equations stated in Art. 8-3 the one that best fits the conditions specified.

4. A 1.05-in. o.d. steel pipe 4 ft long is suspended horizontally in a room in which the air temperature is 40°F. If the outside surface of the pipe is maintained at a temperature of 200°F, how much heat, in Btu/hr, will be transferred to the air by free convection? Neglect the effect of the ends of the pipe.

5. The lower surface of a 3- by 5-ft horizontal plate at a temperature of 330°F is exposed to air at 70°F. Determine the lower surface coefficient of free convection in Btu/(hr)(sq ft)(°F).

6. Air at a pressure of 29.4 psia and at a temperature of 100°F is heated by an 8- by 10-ft horizontal plate facing upward. The upper surface temperature of the plate is 220°F. Determine the surface coefficient of free convection in Btu/(hr)(sq ft)(°F).

7. A tank used for quenching purposes is filled with oil maintained at a temperature of 120°F by means of a submerged wrought-iron pipe coil of 1¼ in. nominal diameter (1.66 in. o.d.). Steam condensing in the coil maintains a coil surface temperature of 220°F. Assuming the characteristics of the oil to be similar to those of transformer oil, determine the surface coefficient of free convection in Btu/(hr)(sq ft)(°F).

8. A tank of water is heated by an electrically heated thin plate, 4 by 6 ft, suspended in a horizontal position in the water. If the plate is maintained at 170°F and the average temperature of the water is 150°F, what is the amount of heat exchanged in Btu/hr?

9. An electrically heated solid cylinder, 6 in. o.d. and 12 in. long, is suspended horizontally in a large tank of water. Determine the watts input to maintain the cylinder at a surface temperature of 200°F when the water is at a temperature of 100°F.

10. Electric windings in the form of a 6-in. cube are immersed in a tank of transformer oil. If the average temperatures of the surfaces of the windings and of the oil are 190 and 150°F, respectively, what must be the power loss in watts?

REFERENCES

1. Saunders, O. A.: *Proc. Roy. Soc. (London)*, **A, 157,** 278–291 (1936).
2. King, W. J.: The Basic Laws and Data of Heat Transmission, *Mech. Eng.*, **54,** 550 (May, 1932).
3. Boelter, L. M. K., and V. H. Cherry: Measurement of Heat Transfer by Free Convection from Cylindrical Bodies by the Schlieren Method, *Trans. ASHVE*, **44,** 491–512 (1938).

CHAPTER 9

HEAT TRANSFER TO BOILING LIQUIDS

9-1. General. Heat transfer to boiling liquids occurs in several kinds of heat-exchange apparatus. The most common examples are the steam boiler and the evaporators of refrigerating plants and chemical industries.

In spite of the wide use of equipment of this sort and notwithstanding the fact that extensive studies have been and are still being made of the heat transfer in this equipment, the fund of information developed concerning heat transfer to boiling liquids is still far from complete or satisfactory.

When a liquid is heated *without* boiling, the heat transfer is governed by the laws of convection and the film conductance can be predicted quite accurately. But when boiling occurs, bubbles of vapor are formed and liberated from the surface in contact with the liquid and turbulence of uncertain magnitude is set up. The film conductance is thereby increased, but in a measure that is influenced by many variable factors. Predictions of the heat transfer by mathematical expressions are therefore uncertain, at least in the light of our present knowledge of the mechanism of vaporization.

Boiling in the ordinary range of temperatures is termed *nucleate* boiling, because of the mechanism by which it takes place through the action of tiny cells of air or other gases that adhere to the heating surface and which serve as *nuclei* for the formation of vapor bubbles. These tiny cells may remain attached to the heating surface when it is first wetted, or they may come from gas dissolved in the liquid.

Researches by Jakob and Fritz[1] have shown that the liquid in direct contact with the heating surface may be superheated several degrees above the saturation temperature corresponding to the pressure, and in some cases, at high rates of heat transfer, even 18 or 19° of superheat have been noted. For the same difference in temperature between the heating surface and the vapor, a rough surface has been found to show less superheating of the liquid and better heat transmission than is shown when the heating surface is smooth. Any contamination of the surface that affects its wettability influences the size and shape of the vapor bubbles, which in turn have an effect upon the heat transfer. The bubbles on a surface that is easily wetted are smaller and are more promptly disengaged from the surface, thereby providing better heat transfer than when the surface is oily or otherwise coated so that its tendency to be wetted is weakened.

It is thought that the adsorption of air or other gases on the heating surface materially affects the film conductance, since tests with freshly cleaned pipe show much greater conductance of heat than when the pipe has been covered with water or other liquid even for a few days.

Other factors that apparently influence the rate of heat transfer during vaporization are (1) a lack of uniformity in the temperature at different points on the heating surface and (2) periodic variations in temperature, which appear to occur in rhythmic fashion. The pressure at which boiling occurs also influences the film conductance. Jakob[2] has shown that for a given liquid and varying pressure the film conductance is proportional to the thermal conductivity of the liquid at saturation temperature, the 0.8 power of the heat flow and the 0.3 power of the surface tension of the bubbles, and is inversely proportional to the kinematic viscosity of the liquid and the 0.3 power of the difference in densities of the liquid and vapor. His expression involving these factors, however, has not yet been verified by sufficient experimental data to warrant the explanation necessary for including it in this text, but it does serve to emphasize the complexity of an accurate calculation of the film conductance for liquids boiling at various pressures even when such factors as the character and condition of the surface are known.

Authorities disagree as to the effect of the size of heating surface upon the evaporation of liquids, but Jakob has presented fairly conclusive evidence that the size exerts little or no effect. He has concluded, however, that the shape and position of the heating surface, since they no doubt affect the disengagement of vapor bubbles, exert some influence upon film conductance.

No difference is apparent in the film conductance of a vertical and a horizontal plane surface; but for a boiler tube, where circulation is assured, Jakob has concluded the value of h_c is of the order of 25 per cent greater than when evaporation takes place on a flat surface.

The results of tests show that when the difference in temperature between the heating surface and the liquid is increased, not only is the flow of heat increased, but because of more violent ebullition, a marked increase occurs in the heat flow per degree of temperature difference between the surface and the vapor, or, in other words, in the film conductance. In fact, in the range of usual heating rates, the film conductance has been found to vary as the 2.4 or the 2.5 power of the temperature difference. Experimental work by Sauer, Cooper, Akin, and McAdams[3] has shown a film conductance h_c for the boiling of water at atmospheric pressure on a freshly cleaned copper tube as high as 5,000 Btu/(hr)(sq ft)(°F). From the work of Jakob, however, it is doubtful if the value of h_c for the film on a tube in continuous service ever exceeds 2,000 Btu/(hr)(sq ft)(°F).

A limit to the increase in film conductance with increase in the heating rate should be expected, for with increased rate of formation of vapor bubbles the heating surface might eventually be covered not by separate bubbles but by a continuous vapor film. That such a limit actually exists was demonstrated by Nukiyama,[4] who boiled water at atmospheric pressure by submerging an electrically heated platinum wire in a horizontal position in a water tank. As the electric input to the wire was increased, the wire temperature was steadily raised to about 300°F. This temperature, however, was a critical point, for with no further increase in the electric input and actually with a slight decrease, the wire temperature suddenly jumped to about 570°F. Further increase in the electric input rapidly brought the wire temperature to more than 1800°F. Subsequent investigations by Sauer, Cooper, Akin, and McAdams,[3] and by others, have shown that at the critical temperature purely nucleate boiling ceases; above this temperature the heating surface is at first insulated not only with separate bubbles but also with bubbles joined together into a continuous vapor film. Further increase in the heat input results in the formation of a continuous film of vapor, covering the heating surface—a type of boiling called *film boiling*. The vapor film formed during film boiling presents such an effective barrier to heat transfer as to cause a rapid rise in the temperature of the heating surface. Abnormally high rates of heat transfer can be obtained by film boiling, however, in spite of the reduced film conductance, provided the heating surface is of material that can withstand a high temperature. Although, at one time, film boiling was considered little more than an interesting phenomenon, the increasing desire for high rates of heat transfer in such developments as nuclear-reactor applications has stimulated interest and much needed research in this mode of heat transfer.

9-2. Expressions for the Film Conductance in Nucleate Boiling. The difficulty in formulating a general expression for the film conductance of boiling liquids that will account for all the possible factors is apparent, even when the expression is limited to the type of boiling which ordinarily occurs. Attempts to develop a general expression by the method of dimensional analysis have not been successful, but a number of empirical equations have been devised whereby at least approximate calculations can be made. Most of these expressions apply within some limited range of pressures, temperatures, or rate of heat flow and therefore should be used with caution.

In most of the equations, h_c is expressed as a function either of the temperature difference Δt between the heating surface and the boiling liquid or of the rate of heat transfer q/A. An example of the former type of expression is shown by Eq. (9-1).

$$h_c = a\,\Delta t^n \qquad \text{Btu/(hr)(sq ft)(°F)} \qquad (9\text{-}1)$$

which applies to the boiling of various liquids at atmospheric pressure, and in which a is a constant for each liquid and n is given various values by different experimenters, within the range of 2.4 and 4.0.

Published values of a and n vary so widely and are in such poor agreement with much of the more recent work in this field that they will not be quoted here.

Cryder and Finalborgo[5] have observed that for boiling at different pressures or temperatures the film conductance varies both with the temperature difference Δt between the surface and liquid and with the boiling temperature t. Their results are expressed by the equation

$$\log_{10} h_c = a + 2.5 \log_{10} \Delta t + bt \qquad (9\text{-}2)$$

where a and b = constants shown in Table 9-1 and t = boiling temperature in degrees Fahrenheit.

TABLE 9-1

Liquid	a	b
Water..	−2.05	0.014
Methanol.......................................	−2.23	0.015
Carbon tetrachloride...........................	−2.57	0.012
Normal butanol................................	−4.06	0.014
26.3% by weight glycerol solution................	−2.65	0.015
Kerosene.......................................	−5.15	0.012
10.1% by weight sodium sulfate solution...........	−2.62	0.016
24.2% by weight sodium chloride.................	−3.61	0.017

This equation is of interest in showing comparisons of the film conductances of various liquids and in showing the manner in which h_c varies with the temperature and the temperature difference. It shows values of h_c, however, that are much higher than those obtained by Jakob and Fritz and a number of other observers whose tests were conducted not with freshly cleaned and polished surfaces, such as were used in Cryder's and Finalborgo's tests, but with surfaces more nearly representing the condition that would be obtained in actual proper operation of heat exchangers. Equation (9-2), therefore, should not be used in design calculations unless the values of h_c found by it are reduced by possibly 50 per cent.

Lack of agreement among equations of various observers in which h_c is expressed as a function of Δt very likely is due not only to difference in the condition of the heating surface but also to differences in technique and in means employed for measuring the temperature of the heating surface and the boiling liquid. Better agreement is shown by equations in which h_c is expressed as a function of the rate of heat transfer q/A at a fixed pressure or temperature. This form of equation, however, is subject to the criticism that the two terms h_c and q/A are interdependent, since q/A is the product of h_c and Δt; and for a fixed rate of heat transfer any error in the measurement of Δt produces a corresponding error in the value of h_c.

When h_c and q/A, obtained from tests, are plotted against each other on rectangular coordinate paper, the result is very nearly a straight line, except at extremely low or high rates of heat transfer; accordingly, the film conductance may be expressed by the equation

$$h_c = c + d\,\frac{q}{A} \qquad \text{Btu/(hr)(sq ft)(°F)} \qquad (9\text{-}3)$$

where $c = h$ intercept of the straight-line portion of the graph and $d =$ slope of the line. A graph in this form for the boiling of water at atmospheric pressure on a flat copper surface is shown in Fig. 9-1.

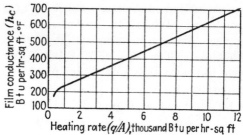

FIG. 9-1. Film conductance for water boiling on a flat copper surface.

Values of c and d for the boiling of various liquids at atmospheric pressure on flat copper surfaces are shown in Table 9-2.

TABLE 9-2

Liquid boiling at atmospheric pressure	c	d
Water	190	0.043
Ammonia	190	0.043
10% sodium sulfate solution	125	0.043
26% glycerol	125	0.043
Kerosene	100	0.030
Methanol	100	0.030
24% sodium chloride solution	100	0 030
Carbon tetrachloride	50	0.025
Normal butanol	50	0.025
Methylene chloride	50	0.025
Methyl formate	50	0.025
Ethyl ether	50	0.025
Methyl chloride (in a clean vertical copper tube)	315	0.037

These values have been obtained from tests by Jakob and Fritz for water and by the relative conductances of various liquids as found by Cryder and Finalborgo. The values for methyl chloride are from work reported by W. H. Martin.[6]

For the boiling of water at atmospheric pressure in boiler tubes the values of c and d (Table 9-2) should be multiplied by about 1.25, or, expressed in equation form,

$$h_c = 240 + 0.054 \frac{q}{A} \qquad \text{Btu/(hr)(sq ft)(°F)} \qquad (9\text{-}4)$$

9-3. Boiling at Various Pressures. Few data are available for showing the effect of evaporating pressure upon the film conductance, although it is well known that at pressures much below atmospheric a reduction in pressure effects a notable decrease in film conductance. When the pressure is increased above atmospheric, the film conductance is at first increased but eventually decreased. The work of Max Jakob and a number of other observers indicates that at a boiler pressure of 225.6 psia the film conductance for water is 1.6 times the conductance at atmospheric pressure but at a pressure of 1,422 psia the conductance is only about 18 per cent greater than at atmospheric pressure.

TABLE 9-3. EFFECT OF EVAPORATION TEMPERATURE
ON FILM CONDUCTANCE

Ratio h/h_a	0.4	0.5	0.6	0.7	0.8	0.9	1.0	1.1	1.2	1.6	1.18
Liquid	Evaporation temperatures										
Water....................	95	127	154	173	188	200	210	220	229	392	589
24% sodium chloride solution.		178	191	204	214	222	229	237			
10% sodium sulfate solution...	125	148	168	180	190	200	211	218	226		
26% glycerol...............		145	168	184	197	207	215	226	237		
Carbon tetrachloride........		91	103	116	136	153	168	180	190		
Methanol.................		66	90	108	125	138	148	156	161		
Kerosene.................			362	380	398	417	428	432			

Table 9-3 has been compiled, mainly from the work of Cryder and Finalborgo,[5] to show the effect of evaporating pressure or corresponding temperature upon film conductance for a number of liquids. For evaporation at boiling points other than those corresponding to atmospheric pressure, the values of h_c obtained by Eqs. (9-3) and (9-4) and Table 9-2 must be multiplied by the values of the ratio h/h_a shown in the table.

Problems

1. Determine, by application of Eq. (9-2), the surface coefficient and the rate of heat transfer for carbon tetrachloride when boiling at atmospheric pressure and at a temperature of 168°F on a freshly cleaned and polished surface that is maintained at a temperature of 188°F.

2. Determine, by application of Eq. (9-3), a design value for the surface coefficient for carbon tetrachloride boiling at atmospheric pressure on a flat copper surface at the same rate of heat transfer as in Prob. 1. The result is what percentage of the value found for h_c in Prob. 1?

3. A vessel is to be designed for evaporating a 26 per cent glycerol solution at a temperature of 168°F by the application of heat to the flat copper bottom of the vessel. To what temperature must the copper be heated in order to transfer 10,000 Btu/(hr)(sq ft of surface)?

4. Heat, at the rate of 7200 Btu/hr, passes through 1 sq ft of inside surface of a boiler tube when boiling water at 229°F. Determine the inside-surface conductance in Btu/(hr)(sq ft)(°F).

5. Water boils in a boiler tube at a temperature of 229°F. If the latent heat of evaporation is 958 Btu/lb and if 5 lb of water is evaporated per hour for each square foot of inner surface, what is the metal temperature at the inner surface of the tube?

6. The pressure in a boiler is such that the water boils at 392°F. Compute (a) the inside-surface conductance in Btu/(hr)(sq ft)(°F) and (b) the metal temperature at the inner surface of the boiler tubes, for a heating rate of 8000 Btu/(hr)(sq ft of inner surface).

7. A 24 per cent sodium chloride solution is to be boiled in a flat-bottomed copper vessel at a temperature of 237°F. Assuming an inside-surface coefficient of 600 Btu/(hr)(sq ft)(°F), determine (a) the rate of heat transfer in Btu/(hr)(sq ft of surface) and (b) the temperature of the copper surface.

8. In the evaporator of a refrigerating machine, ammonia boils at atmospheric pressure (−28°F). If the latent heat of evaporation is 589.3 Btu/lb and if 2 lb of ammonia is evaporated per hour per square foot of inner surface, what would be the metal temperature at the inner surface of the evaporator?

9. Water is boiled at 212°F on a flat surface of thin copper which receives 5000 Btu/(hr)(sq ft) in the form of radiant heat. The source of heat is a Cr-Ni alloy plate of the same size as the copper. The surfaces and their arrangement are such as to have a configuration factor F_A of 0.80 and an emissivity factor F_e of 0.30. What must be the surface temperature of the Cr-Ni plate?

REFERENCES

1. Jakob, M., and W. Fritz: *Forschr. Gebiete Ingenieurw.*, **3**, 161 (1932).
2. Jakob, M.: *Armour Inst. Tech. Tech. Bull.*, vol. 2, no. 1 (July, 1939).
3. Sauer, E. T., H. B. H. Cooper, G. A. Akin, and W. H. McAdams: *Mech. Eng.*, **60**, 669–675 (1938).
4. Nukiyama, S.: *Jr. Soc. Mech. Engrs. (Japan)*, **37**, 366–374, S53–54 (1934).
5. Cryder, D. S., and A. C. Finalborgo: *Trans. Am. Inst. Chem. Engrs.*, **33**, 346–361 (1937).
6. Martin, W. H.: *Refrig. Eng.*, **46**, 175 (September, 1938).

CHAPTER 10

CONDENSING VAPORS

10-1. General. It has already been observed that when heat is added to saturated or superheated vapors or is removed from superheated vapors without sufficient cooling to cause a change of state from gas to liquid, the laws of heat transfer as applied to gases hold true. When the vapor is cooled to the extent where condensation takes place, heat is transferred in a fundamentally different manner. The condensation of vapor liberates a considerable amount of heat, but at the same time a barrier is set up in the form of a liquid film which either partially or completely covers the cooler surface. The thickness of this film is influenced both by its viscosity and by the position and height of the surface. Drainage from a vertical or an inclined surface is naturally more rapid than from a horizontal one, and the film is accordingly thinner; but if the vertical height is great, the accumulation of condensation on the lower portion of the surface will thicken the film and make the lower portion less effective than the upper in transmitting heat. Roughness of the surface also affects the drainage and thickness of the film, a rough surface retaining a thicker film than a smooth one.

In view of the foregoing observations it should be apparent that where tubes are used as condensing surface, they will be most effective when smooth and in a horizontal position with condensation occurring on the outside of the tube where the condensate may drip off readily.

Additional factors that at times may influence the conductance of films formed by condensation are (1) the possible effect of velocity of the vapor over the surface of the film if the velocity is high and (2) the occasional formation of the film as a collection of disconnected liquid drops instead of a continuous film.

A simple demonstration made by bringing a number of different cold surfaces into contact with a jet of steam will usually show a variety of forms in which condensation may appear, particularly when the smoothness of the surface is varied and when some of the surfaces are clean and others are coated with oily substances. In general, however, these various forms may be classified either as continuous-film or as dropwise condensation. Of the two forms, continuous-film condensation is much the more common, although less effective in the matter of heat transmission.

Observations of the manner in which heat from condensation of vapors is transmitted were developed in mathematical form by Nusselt[1] in 1916. In the development he assumed the vapor to condense in the form of a continuous film flowing in streamline motion down the cooling surface. Gravity alone was considered responsible for the downward movement, and no allowance was made for the possible effect of the velocity of the vapor flowing over the surface of the film.

Nusselt's theory has proved to be in fair agreement with most of the experimental data that have been obtained, although it has been observed that Nusselt's equations [(10-1) to (10-4)] generally present results that are slightly low, or at least conservative.

10-2. Horizontal Tubes. Nusselt's equation for the conductance of a continuous film formed by the condensation of a pure saturated vapor on the outside of a single horizontal tube is

$$h_c = 0.725 \left(\frac{k^3 \rho^2 g r}{D \mu \, \Delta t} \right)^{1/4} \tag{10-1}$$

If the value of g (4.17×10^8 ft/hr^2) is moved outside the brackets, the equation may be expressed as

$$h_c = 103.7 \left(\frac{k^3 \rho^2 r}{D \mu \, \Delta t} \right)^{1/4} \tag{10-2}$$

where, in both equations, r = latent heat of vaporization, Btu/lb$_m$ at the saturation temperature of the vapor.

Δt = difference in temperature of the vapor and the wall, °F.

D = outside diameter of the tube, ft.

All other values, which apply to the liquid film, follow standard nomenclature (shown in Table A, page xiv), and are evaluated at the arithmetic mean of the vapor and wall temperatures.

Equations (10-1) and (10-2) may be applied with little error to a bank of horizontal tubes if D is made equal to the sum of the outside diameters of the tubes in a vertical row. In this case it is assumed that the increased thickness of film on the lower tubes due to the accumulation of drip from the upper rows is at least partially offset by the agitation caused by the drip as it falls from one tube to another.

10-3. Vertical and Inclined Surfaces. The film conductance for pure saturated vapors condensing on vertical and inclined surfaces has been expressed by Nusselt as

$$h_c = 0.943 \left[\frac{k^3 \rho^2 g r (\sin \psi)}{L \mu \, \Delta t} \right]^{\frac{1}{4}} \tag{10-3}$$

or if the value of g $(4.17 \times 10^8$ ft/hr^2) is moved outside the brackets, the equation becomes

$$h_c = 134.8 \left[\frac{k^3 \rho^2 r (\sin \psi)}{L \mu \, \Delta t} \right]^{\frac{1}{4}} \tag{10-4}$$

where L = slant height of the surface in feet and ψ = angle between the surface and the horizontal. r = latent heat of vaporization, Btu/lb$_m$ at the saturation temperature of the vapor. All other units follow standard nomenclature (shown in Table A, page xiv), evaluated at the arithmetic mean of the vapor and wall temperatures.

Equations (10-3) and (10-4) may be used as approximations for vertical pipes. The term $\sin \psi$ need be included for inclined surfaces only, since for vertical surfaces $\sin \psi = 1$.

Where repeated calculations are to be made of the film conductance for condensation of various pure saturated vapors, alignment charts as devised by Chilton, Colburn, Genereaux, and Vernon[2] will be found convenient.

A comparison of Eqs. (10-2) and (10-4) for horizontal and vertical surfaces shows the value of the constant 134.8 in Eq. (10-4) to be higher than the constant in Eq. (10-2), viz., 103.7. This comparison should not lead to the conclusion that the film conductance is greater for the vertical than for the horizontal surface, for, it should be noted, the dimension L, the height of the vertical surface, which appears in the denominator, is ordinarily vastly greater than the corresponding dimension D, the diameter of the horizontal pipe.

The following numerical example will show the advantage in increased film conductance when a tube of ordinary dimensions used to condense steam is placed in the horizontal rather than the vertical position.

Example 10-1.　Compare the outside-film conductances of a 1-in. tube 5 ft long, in the horizontal and vertical positions, when it is used to condense dry saturated steam at atmospheric pressure.　Assume a tube-wall temperature of 200°F.

At the average water-film temperature of $(212 + 200)/2 = 206$, the values of the terms to be substituted in Eqs. (10-2) and (10-4) are

$$k = 0.393 \qquad \mu = 0.71$$
$$\rho = 60.0 \qquad D = 0.0833$$
$$r = 970.3 \qquad L = 5$$
$$\Delta t = 12$$

By Eq. (10-2) the film conductance for the horizontal tube is

$$h_c = 103.7 \left(\frac{0.393^3 \times 60.0^2 \times 970.3}{0.0833 \times 0.71 \times 12} \right)^{1/4}$$
$$= 103.7(299,000)^{1/4} = 103.7 \times 23.4$$
$$= 2426 \text{ Btu}/(\text{hr})(\text{sq ft})(°\text{F})$$

By Eq. (10-4) the film conductance for the vertical tube is

$$h_c = 134.8 \left(\frac{0.393^3 \times 60.0^2 \times 970.3}{5 \times 0.71 \times 12} \right)^{1/4}$$
$$= 134.8(4,983)^{1/4} = 134.8 \times 8.4$$
$$= 1130 \text{ Btu}/(\text{hr})(\text{sq ft})(°\text{F})$$

The horizontal position, therefore, in its influence upon the film conductance, is $2426/1130$, or 2.14 times as effective as the vertical.

A peculiarity that may be noted in Eqs. (10-1) to (10-4) is that the value of h_c decreases when the temperature difference Δt increases.　This decrease in h_c is in keeping with the increase in resistance caused by the building up of the film when, because of a large temperature difference, the rate of condensation is increased.　A review of the equations presented for forced and for free convection will show that for any certain film temperature with forced convection of all fluids except those of very high viscosity, the film conductance is not affected by the temperature difference, but for free convection h_c increases with an increase in Δt.　Tests in which *dropwise* condensation has appeared have shown an increase in the film conductance when the temperature difference has been increased, indicating no building up of the film but, instead, a more rapid removal of the drops.

10-4. Dropwise Condensation. In many of the early investigations of vapor condensation erratic or unanticipated results occurred which are now explainable by the assumption that the condensation may have been deposited in dropwise and not in film form.

In laboratory experiments[3] at the Massachusetts Institute of Technology in which the condensing surface was made visible through pyrex glass, film conductances for steam condensation of the order of 14,000 Btu/(hr)(sq ft)(°F) have been found for dropwise condensation as compared with a value of the order of 2000 for continuous-film condensation.

Numerous tests[4] at the Massachusetts Institute of Technology have shown that dropwise condensation of steam is much more likely to occur on smooth than on rough surfaces. Dropwise condensation does not occur unless the cooling surface is in some way contaminated, but a too heavy accumulation of contaminant may promote the formation of a continuous film. An almost infinitesimally small amount of contaminating substance may induce dropwise condensation on some metals but not on others. Some contaminants seem to depend for their activity upon the amount of noncondensable gas present. In order to be effective the contaminant must adhere firmly to the surface. Certain of the fatty acids have been found generally effective, especially so when the condensing surface is of copper, brass, nickel, or chromium plate.

The use of added agents to control the type of condensation of vapors is covered by a United States patent granted to W. M. Nagle.

Since dropwise condensation can be expected only under carefully controlled conditions, film condensation should ordinarily be assumed in the design of condensing equipment.

10-5. Effect of Noncondensable Gas. The presence of air or other noncondensable gas mixed with a vapor is to be avoided if high rates of heat transfer by condensation are desired, except in the case cited, where a small amount of noncondensable gas appears to increase the effectiveness of some contaminants in promoting dropwise condensation. In tests of steam radiators and other steam-heating equipment in which steam is condensed, the rate of condensation will be notably reduced if even the small amount of air ordinarily present in steam is not thoroughly vented.

The effects of various percentages of air upon the film conductance of steam condensing on a horizontal 3-in. tube have been reported by Othmer.[5] The results are dependent both upon steam temperature and upon the difference in temperature of the steam and the tube surface; but the effect of the presence of air may be well illustrated by a specific instance where, with steam at 230°F and a temperature difference between the vapor and the tube of 10°, an air content of 4.53 per cent (by volume) reduced the film conductance from 2800 to 480 Btu/(hr)(sq ft)(°F).

It has already been noted that when the film conductance for a pure saturated vapor condensing on surfaces in different positions is computed by the applications of Eqs. (10-1) to (10-4), the result is a conservative figure. The presence of noncondensable gas may lower the film conductance, but with proper precautions the presence of such gas may ordinarily be prevented. It is more likely that the conductance may be raised because of turbulence created by vapor velocity or because of the existence of dropwise condensation.

10-6. Approximate Values of Conductance for Various Condensing Surfaces. Where the conditions of operation of condensing equipment cannot be predicted with certainty, the use of approximate or representative values of film conductance is justified, certainly in preference to the use of average over-all coefficients of heat transfer. For such purposes, the film conductance for steam condensing on horizontal tubes is usually estimated at 2000 Btu/(hr)(sq ft)(°F). For steam condensing on inner surfaces, as in steam radiators and fan coils, a value of the order of 1200 is considered more representative, owing to the slower drainage of condensate from such surfaces. From tests of ammonia condensers, Kratz, Macintire, and Gould[6] have computed the conductance of the ammonia film to be 1635 Btu/(hr)(sq ft)(°F). Corresponding values from various sources for several other condensing vapors are as follows:

Carbon tetrachloride........... 300
Benzene...................... 350
Diphenyl vapor............... 300

It is well to keep in mind the comparison of these values with representative values for the film conductances of the most common fluids: water, air, and steam. The illustrative examples 7-1 and 8-2 show film conductances for water in forced and free convection of 686 and 200 Btu/(hr)(sq ft)(°F), respectively. Corresponding values for air, as shown in Exs. 7-3 and 8-1, are only 3.56 and 0.97 Btu. Example 7-2 shows a value for the film conductance of superheated steam of 133 Btu/(hr) (sq ft)(°F), which is not far below the value for water in free convection but much below the values of 2000 and 14,000 quoted earlier in this chapter for steam when condensing in continuous film and in dropwise formations.

10-7. Dehumidification. Experimental results have corroborated the theory that when a mixture of a condensable vapor and a noncondensable gas, such as a mixture of water vapor and air, is brought into contact with a surface colder than the dew point of the mixture, the removal of some of the vapor by condensation leaves a layer of gas next to the liquid film in which there is less concentration of vapor than in the main body of the mixture. The conversion of vapor into condensate lowers the partial pressure of the vapor at the surface of the film of condensate and sets up a continuous flow of vapor from the main body of the mixture through the gas film. Thus, the rate of heat transfer by condensation is not the same as when a pure vapor is condensed, for the presence of vapor at the cold surface is governed by the rate at which it can be diffused through the gas film.

No mathematical analysis of this process will be included here, and further discussion will be limited to the most common application of this process, viz., in the dehumidification of air for air conditioning, which will be discussed in a later chapter.

Problems

1. Compute the film conductance for pure saturated methyl chloride at 85.3 psia and 80°F, condensing on the outside of 1¼-in. horizontal wrought-iron pipes (1.66 in. o.d.) when the pipes are cooled to an average outside-surface temperature of 70°F. The density of the liquid methyl chloride is 57.0 lb_m/cu ft, and the latent heat of vaporization is 161.9 Btu/lb. There are three pipes in each vertical row.

2. Compute the film conductance for pure saturated sulfur dioxide at 63 psia and 83°F, condensing on a surface 10 in. high, inclined at an angle

of 60 deg with the horizontal. The average surface temperature is 75°F.
The density of the liquid sulfur dioxide is 85.1 lb$_m$/cu ft, and the latent heat
of vaporization is 145.6 Btu/lb.

3. Select the correct values in the following list:

a. The surface coefficient for condensing steam on horizontal tubes is
commonly of the order of 2; 20; 200; 2000; 20,000 Btu/(hr)(sq ft)(°F).

b. The inside-surface coefficient for superheated steam in a steam header
is commonly of the order of 1; 10; 100; 1000; 10,000 Btu/(hr)(sq ft)(°F).

c. The surface coefficient for boiling water at atmospheric pressure is
commonly of the order of 2; 20; 200; 2000; 20,000 Btu/(hr)(sq ft)(°F).

d. The surface coefficient of free convection for indoor air is commonly
of the order of 1; 10; 100; 1000; 10,000 Btu/(hr)(sq ft)(°F).

4. Heat is removed at the rate of 940 Btu/hr from the wall of a 1-in. i.d.
tube 36 in. long in which carbon tetrachloride vapor is condensed at a saturation temperature of 170°F. What must be the temperature of the inner
surface of the tube?

5. What length of horizontal 1-in. o.d. tube at a temperature of 148°F is
required to condense 20 lb/hr of dry saturated steam at 152°F on its outer
surface?

REFERENCES

1. Nusselt, W.: *Z. Ver. deut. Ing.*, **60**, 541, 569 (1916).
2. Chilton, T. H., A. P. Colburn, R. P. Genereaux, and H. C. Vernon:
   Trans. ASME, Petroleum Division, **55**, 1933.
3. Nagle, W. M., G. S. Bays, Jr., L. M. Blenderman, and T. B. Drew:
   Trans. Am. Inst. Chem. Engrs., **31**, 593 (1935).
4. Drew, T. B., W. M. Nagle, and W. Q. Smith: The Conditions for Dropwise Condensation of Steam, *Trans. Am. Inst. Chem. Engrs.*, **31**, 605
   (1935).
5. Othmer, D. F., *Ind. Eng. Chem.*, **21**, 576 (1929).
6. Kratz, A. P., H. J. Macintire, and R. E. Gould, *Univ. Ill. Eng. Expt. Sta.
   Bull.* 209, 1930.

CHAPTER 11

OVER-ALL TRANSFER OF HEAT

11-1. General. In the study of heat transfer up to this point, conduction, radiation, and convection have been considered separately. Heat is usually transferred, however, from one medium to another not by one mode alone but by two or all three of the modes. In one of the simplest cases, for example, where heat is transferred from a uniformly heated room through a solid wall to the outdoor air, the heat is conveyed to the inner surface of the wall mainly by convection but also by radiation from objects at room temperature to the cooler surface of the wall. Heat then flows through the wall by conduction and finally leaves the outer surface by convection and radiation. For steady-flow conditions, the same amount of heat is transmitted through each of the three successive barriers, but the drops in temperature across the wall and across the surface films may be very different. Three different amounts of heat may therefore be expected per degree of temperature change across each of the three barriers; or otherwise stated, there may be no similarity among the conductances of the three barriers.

It is often desirable to express the heat transfer not as the amount per degree of change in temperature across a single barrier but as the amount per degree of over-all difference in temperature of the hotter and cooler mediums. This is the case especially where measurement of surface temperatures is impracticable. The amount of heat transferred in Btu/(hr)(sq ft of surface)(°F difference in temperatures of the two mediums) is known as the over-all coefficient of heat transmission and is expressed by the symbol U.

In dealing with the over-all coefficient U, one should not lose sight of the practical significance of the individual factors affecting the flow of heat.

11-2. Practical Significance of Factors Affecting Flow of Heat.

The flow of heat from a region of higher to one of lower temperature is analogous to the flow of a fluid from a higher to a lower level. The flow of a fluid may be restricted by a dam or weir or valve or any other form of obstruction. Similarly, the flow of heat may be restricted by a material of low conductivity, by a sluggish or viscous film, or by a surface with low radiating ability.

If the flow of water from a millpond is to be increased, widening or deepening the channel will ordinarily have slight effect, but removal of the obstructing dam may produce so great an effect as to create a flood. Similarly, if the flow of heat between two temperature levels is to be increased, the most productive change that can be made will be that which will reduce the most effective barrier. In some cases that barrier may be caused by low conductivity; in others it may be because of low radiating value; and in numerous instances it may be the lack of effective convection.

One of the most important observations that can be made with reference to the factors affecting the flow of heat is that if the individual barriers differ greatly in effectiveness, only that one which offers the greatest resistance will have much effect upon the transmission. If the relative importance of the individual resistances is overlooked, very erroneous conclusions may be drawn. As an illustration, salesmen for warm-air furnaces of thin metal construction often claim an advantage for their furnace in greater heat transfer because of the thinness of the metal. To a student of heat transfer who is aware that in this case the resistance to conduction through the metal is insignificant in comparison with the surface resistances, this claim would bear little or no weight. (This claim should not be confused with the justifiable claims of quick response and low chimney losses that may be obtained for intermittent heating with lightweight furnaces.)

Again, in warm-air furnace systems of heating, for many years the basement warm-air pipes were completely covered with one or two thicknesses of asbestos paper, not only for the purpose of sealing the joints but also supposedly for decreasing the loss of heat from the pipes. This practice was improved upon by limiting the use of the asbestos paper to narrow strips for covering the joints or by replacing the paper by an air-cell covering only when the results of research had shown that the high radiating value of paper, in comparison with a bright metal surface, more than offsets its advantage of low conductivity.

Numerous illustrations can be cited of improvements in the design of mechanical equipment that have resulted from application of the principles of heat transfer. In practically all instances, however, these improvements were made possible only when the relative importance of the factors affecting the transfer of heat was recognized.

An expression will now be developed for the over-all coefficient U in terms of the various resistances to heat transfer. This expression not only will serve for the calculation of U but in any specific instance will show also, by numerical values, the relative importance of the various resistances encountered.

11-3. Calculation of Over-all Coefficient of Transmission. Figure 11-1 shows a temperature gradient applying to the transfer of heat from room air at a temperature t_1, through an inside-sur-

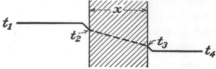

FIG. 11-1. Solid wall.

face resistance to a wall surface at temperature t_2, then through the solid wall to the outer surface at temperature t_3, and finally through an outside-surface resistance to outdoor air at temperature t_4.

The flow of heat through unit area of the surface resistance on the room side of the wall is

$$\frac{q_1}{A} = h_i(t_1 - t_2)$$

where h_i = inside-surface conductance. The flow of heat by conduction through the same area of the wall is

$$\frac{q_2}{A} = \frac{k}{x}(t_2 - t_3)$$

where k = thermal conductivity of the wall material and x = wall thickness. The flow through the same area of the outer-surface resistance is

$$\frac{q_3}{A} = h_o(t_3 - t_4)$$

where h_o = outside-surface conductance. But for steady-flow conditions, q_1/A, q_2/A, and q_3/A are all equal and may be expressed as

$$\frac{q}{A} = U(t_1 - t_4) \qquad (11\text{-}1)$$

where U = over-all coefficient of transmission. The over-all coefficient U may be expressed in terms of the three successive barriers by the following procedure:

$$U(t_1 - t_4) = h_i(t_1 - t_2)$$
$$= \frac{k}{x}(t_2 - t_3)$$
$$= h_o(t_3 - t_4)$$

Transposing,

$$\frac{U}{h_i}(t_1 - t_4) = t_1 - t_2$$

$$\frac{Ux}{k}(t_1 - t_4) = t_2 - t_3$$

$$\frac{U}{h_o}(t_1 - t_4) = t_3 - t_4$$

Adding,

$$U\left(\frac{1}{h_i} + \frac{x}{k} + \frac{1}{h_o}\right)(t_1 - t_4) = t_1 - t_4$$

$$U = \frac{1}{\dfrac{1}{h_i} + \dfrac{x}{k} + \dfrac{1}{h_o}} \qquad \text{Btu/(hr)(sq ft)(°F)} \qquad (11\text{-}2)$$

The over-all coefficient U can thus be computed for any wall or other flat barrier of thickness x by substitution of the appropriate values of h_i, h_o, k, and x in Eq. (11-2).

It should be observed that since the *conductances* of the two films and the wall are h_i, h_o, and k/x, their reciprocals, $1/h_i$, $1/h_o$ and x/k, which appear in the denominator of Eq. (11-2), represent the three *resistances* to the flow of heat.

In cases where more than three successive resistances are encountered, Eq. (11-2) may be amplified by the addition, in the denominator, of the fractions that represent the additional resistances. Thus in the case shown in Fig. 11-2 for a wall consisting of two materials of thicknesses x_1 and x_2 and con-

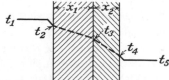

Fig. 11-2. Composite wall.

ductivities k_1 and k_2, the over-all coefficient will be expressed as

$$U = \frac{1}{\dfrac{1}{h_i} + \dfrac{x_1}{k_1} + \dfrac{x_2}{k_2} + \dfrac{1}{h_o}} \qquad (11\text{-}3)$$

When air spaces form a part of wall construction, an additional resistance at each added surface is introduced. Because of the difference in temperature on the two sides of the air space, some degree of air movement is quite sure to be set up within the space unless the space is unusually narrow. Heat is transferred across the space by convection currents and by radiation; but because of the circulation of the air, no conduction can take place aside from that which is inherently involved in the process of convection. The resistance of an air space may therefore be expressed approximately as the sum of the two surface resistances, or as $2/h_i$, with no x/k fraction applying to the air space included in the expression for U.

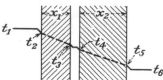

FIG. 11-3. Wall with air space.

Tests of the thermal resistances of narrow vertical air spaces, such as may be used in wall construction, have shown that for air spaces less than about $1\frac{1}{2}$ in. in width, the conductance is a function both of the width of the space and of the mean temperature of the air. Accordingly, the preferred procedure for computing the value of U for a wall such as is shown in Fig. 11-3, in which there is an air space less than $1\frac{1}{2}$ in. in width, is to apply the expression in the form

$$U = \frac{1}{\dfrac{1}{h_i} + \dfrac{x_1}{k_1} + \dfrac{1}{C} + \dfrac{x_2}{k_2} + \dfrac{1}{h_o}} \qquad \text{Btu/(hr)(sq ft)(°F)} \quad (11\text{-}4)$$

where C = conductance of the air space as shown by Fig. A-11 (from the work of Rowley and Algren[1]).

Calculations by Eq. (11-4) produce slightly higher values of U than are found by the former method, but the result is undoubtedly more accurate in the case of construction with air spaces of the widths investigated, viz., $\frac{1}{8}$ to $1\frac{1}{2}$ in. Application of the former method is therefore restricted usually to construction with air spaces wider than $1\frac{1}{2}$ in.

In the application of the foregoing expressions for the over-all coefficient to specific problems, the values of h_i and h_o may be computed as the sum of the conductances by convection and radiation in accordance with the methods outlined in the chapters on convection and radiation. In the case of calculations of over-all coefficients for walls of buildings, where the temperatures ordinarily do not vary through a wide range, the usual procedure is to assign a value for the inside-surface conductance h_i ranging from 1.65 to 2 Btu, increasing in proportion to the roughness of the surface, and to evaluate the outside-surface conductance h_o according to the outdoor air velocity, both in accordance with the data shown in Fig. A-10. In problems dealing with the calculation of the heating requirements of buildings, the air velocity to be encountered in coldest weather is most commonly assumed to be 15 miles/hr, for which the value of h_o, applying to common building materials, as shown in Fig. A-10, ranges from 5 to 9 Btu/(hr)(sq ft)(°F).

In textbooks and reference books on the subject of heating and ventilating, where data are presented for computing values of U, the thickness of materials is usually expressed in *inches*, and values shown for thermal conductivity k appear in terms of Btu/(hr)(sq ft)(°F) for 1-*in.* and not for 1-ft thickness. No difference results in the calculated values of U so long as the dimension x is expressed in the same units as the thickness upon which the conductivity is based. The difference in the units employed in heating and ventilating and in heat-transfer texts, however, should be observed in order to avoid possible confusion in problems of conduction of heat.

The following examples will show the application of Eqs. (11-3) and (11-4).

Example 11-1.   Determine the over-all coefficient of transmission U for a wall of 12-in. brick, covered on the inside with ¾ in. of plaster. Assume an outdoor wind velocity of 15 mph. k_1 for plaster $= 0.27$ Btu/(hr)(sq ft) (°F/ft); k_2 for 8-in. common brick $= 0.40$ and k_3 for 4-in. face brick $=0.76$; $h_c = 1.7$; $h_o = 7.2$ (Fig. A-10).

$$U = \cfrac{1}{\cfrac{1}{1.7} + \cfrac{0.75}{12 \times 0.27} + \cfrac{8}{12 \times 0.40} + \cfrac{4}{12 \times 0.76} + \cfrac{1}{7.2}}$$

$$= \frac{1}{0.588 + 0.231 + 1.667 + 0.439 + 0.139} = \frac{1}{3.064}$$

$$= 0.326 \text{ Btu/(hr)(sq ft)(°F)}$$

It should be observed that a comparison of the five resistances can be made from the five values 0.588, 0.231, 1.667, 0.439, and 0.139.

Example 11-2.   Determine the over-all coefficient of transmission U for the wall of Ex. 11-1, but with an air space of $1\frac{1}{2}$ in. between the brick and the plaster.   (Neglect the resistance of the furring strips and lath required for supporting the plaster.)   Values of k_1, k_2, k_3, h_c, and h_o are as in Ex. 11-1. C for the air space at an assumed mean temperature of $40° = 1.1$ (Fig. A-11).

$$U = \cfrac{1}{\cfrac{1}{1.7} + \cfrac{0.75}{12 \times 0.27} + \cfrac{1}{1.1} + \cfrac{8}{12 \times 0.40} + \cfrac{4}{12 \times 0.76} + \cfrac{1}{7.2}}$$

$$= \frac{1}{0.588 + 0.231 + 0.909 + 1.667 + 0.439 + 0.139} = \frac{1}{3.973}$$

$$= 0.252 \text{ Btu/(hr)(sq ft)(°F)}$$

11-4. Over-all Coefficients of Transmission.

Values of U have been computed and published for all ordinary types of building construction comprising a great variety of materials of various thicknesses.   The most complete tabulation of these values is found in the Guide of the American Society of Heating and Air Conditioning Engineers, published annually by the Society. These coefficients have been computed on the assumption of dry materials.   A few representative values are shown in Table 11-1.

TABLE 11-1. COEFFICIENTS OF HEAT TRANSMISSION FOR VARIOUS FORMS OF BUILDING CONSTRUCTION
(Based on an outside wind velocity of 15 miles/hr)

$$\frac{U}{\dfrac{Btu}{(hr)(sq\ ft)(°F)}}$$

Walls:
Brick wall, 8 in. thick, plain	0.50
Brick wall, 12 in. thick, plain	0.36
Brick wall, 8 in. thick, $\frac{1}{2}$ in. plaster	0.46
Brick wall, 8 in. thick, plaster on wood lath, furred	0.30
Hollow tile, 8 in. thick, stucco exterior	0.40
Limestone or sandstone, 8 in. thick, plain	0.71
Concrete, monolithic, 10 in. thick	0.62
Frame wall, clapboard, 1-in. wood sheathing, lath and plaster	0.25
Frame wall, clapboard, $2\frac{5}{32}$ in. rigid insulation, lath and plaster	0.19
Frame wall, clapboard, 1-in. wood sheathing, lath and plaster, $3\frac{5}{8}$-in. rock wool fill between studding	0.07

Ceilings and roofs:
Wood lath and plaster, no floor above	0.62
Wood lath and plaster, joists, 1-in. pine floor above	0.28
Metal lath and plaster, $3\frac{5}{8}$-in. rock wool fill between joists, no floor above	0.08
Wood shingles on wood strips, rafters, no ceiling	0.46

Windows:
Ordinary windows	1.13
Double windows	0.45

Although most of the published values of U for building construction have been computed by the method just outlined, a sufficient number of them have been verified experimentally to give assurance that the computed values are satisfactorily accurate for design purposes.

Over-all coefficients of heat transmission for metal tubes as applied in a variety of forms of power-plant and refrigeration apparatus are shown in Table 11-2. Since these coefficients are

TABLE 11-2. OVER-ALL COEFFICIENTS FOR METAL TUBES IN
POWER-PLANT AND REFRIGERATION APPARATUS

	U $\dfrac{Btu}{(hr)(sq\,ft)(°F)}$
Steam boiler tubes..	3–14
Economizers..	2–10
Steam superheaters, close to furnace......................	8–15
Surface condenser (steam).................................	200–1000
Feed-water heaters...	200–1500
Air preheaters..	1–4
Ammonia condensers:	
Double pipe...	150–250
Shell and tube..	150–300
Atmospheric..	60–200
Brine coolers:	
Double pipe...	150–300
Shell and tube..	90–100
Water cooler, shell and coil..............................	15–25
Cooling coils:	
Water to air in unit coolers.............................	5–9
Boiling refrigerant to air in unit coolers................	4–8
Brine to unagitated air..................................	$2–2\frac{1}{2}$

dependent upon such variable factors as the velocities and temperatures of the fluids and upon the cleanliness of the surfaces, it is obvious that the tabulated values can be no more than rough approximations showing only the range in which they may usually be found. Where the factors affecting the over-all coefficients are known or specified, the coefficients should be calculated by the methods already outlined.

11-5. Experimental Determination of the Over-all Coefficient U. The most common method of determining the over-all coefficient U experimentally is by use of the guarded hot box. The apparatus as standardized by the American Society of Heating and Air Conditioning Engineers[2] consists of two cubical boxes of 3-

and 5-ft inside dimensions. Each box is made of five well-insulated walls and with one end open. The smaller box is located with uniform spacing inside the larger and with the open ends of both boxes flush against the specimen of wall section that is to be tested. The same temperature is maintained in both boxes by means of nonluminous electric heaters; and uniformity of temperatures at all elevations within the boxes is maintained by fans arranged to circulate the air gently. In a test with this apparatus, since no difference exists in the temperature of the air in the inner and outer boxes, all the electric input to the heaters and fans within the inner box represents heat that can flow only through the adjacent 3- by 3-ft section of the test specimen. The coefficient U is accordingly established as the heat equivalent of the electric input to the inner box divided by the 9 sq ft of specimen surface and by the measured difference in temperature of the air on the two sides of the test specimen.

The Nicholls heat meter[3] is another device for determining the over-all coefficient U. This device is portable and especially convenient for measuring the transmission through actual building walls in place. The apparatus consists of a plate of Bakelite ⅛ in. thick and 2 ft square, firmly clamped to the test wall. Thermocouples embedded flush with the surfaces of the plate are employed to measure the difference in temperatures of the two surfaces. The design of this device is based upon the principle that the temperature drop across any barrier of known thickness and conductivity, such as the Bakelite plate, is a measure of the heat flow. Although the Bakelite plate itself interposes a resistance to flow through the test wall that is not normally present, its effect is comparatively small. The instrument meter is calibrated against a guarded hot plate and is graduated to indicate directly the hourly quantity of heat flow. U is calculated as the heat flow per hour divided by the area of the surface of the wall in contact with the Bakelite plate and by the measured difference in the temperatures of the air adjacent to the two sides of the wall.

11-6. Measurement of Surface. The foregoing considerations of the value of U have applied to plane-surfaced walls where the extent of surfaces on both sides is the same. Where heat is transmitted through thick cylindrical walls or thick-walled pipes, there is an appreciable difference in the extent of inner and outer surfaces. For steady flow of heat through the inner and outer surfaces A_i and A_o, for which the over-all coefficients are U_i and U_o, respectively, A_iU_i must equal A_oU_o. U_i and U_o must differ in inverse ratio of A_i to A_o, and the value of U therefore must be applied to a square foot of inner or of outer surface as may be specified.

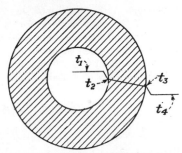

FIG. 11-4. Thick-walled cylinder.

An expression for the outside-surface coefficient U_o will now be derived by consideration of the heat transfer through a 1-ft length of cylindrical wall with temperatures as denoted in Fig. 11-4.

Equating the over-all heat transfer to the transfer through the three successive barriers,

$$A_oU_o(t_1 - t_4) = h_i(t_1 - t_2)A_i$$
$$= \frac{2\pi k_m(t_2 - t_3)}{\ln (r_3/r_2)}$$
$$= h_o(t_3 - t_4)A_o$$

Transposing,

$$\frac{A_oU_o}{A_ih_i}(t_1 - t_4) = t_1 - t_2$$

$$\frac{A_oU_o \ln (r_3/r_2)}{2\pi k_m}(t_1 - t_4) = t_2 - t_3$$

$$\frac{A_oU_o}{A_oh_o}(t_1 - t_4) = t_3 - t_4$$

Adding,

$$U_o\left(\frac{A_o}{A_ih_i} + \frac{A_o \ln (r_3/r_2)}{2\pi k_m} + \frac{1}{h_o}\right)(t_1 - t_4) = t_1 - t_4$$

or

$$U_o = \frac{1}{\dfrac{A_o}{A_ih_i} + \dfrac{A_o \ln (r_3/r_2)}{2\pi k_m} + \dfrac{1}{h_o}}$$

or in terms of the outer and inner diameters (replacing A by πD),

$$U_o = \frac{1}{\dfrac{D_o}{D_ih_i} + \dfrac{D_o \ln (D_o/D_i)}{2k_m} + \dfrac{1}{h_o}} \qquad \text{Btu/(hr)(sq ft)(°F)} \quad (11\text{-}5)$$

The following example will illustrate the application of Eq. (11-5).

Example 11-3.   Calculate the over-all coefficient of transmission U, in terms of the outer surface, for an extra-heavy 8-in. horizontal steel pipe through which superheated steam at 250 psia and 600°F flows at a velocity of 10,000 fpm when the outer surface of the pipe is exposed to air at 70°F. The pipe is located in a large room.

Solution.   By Eq. (11-5)

$$U_o = \frac{1}{(D_o/D_i h_i) + (D_o/2k_m)\ln(D_o/D_i) + (1/h_o)}$$

$$D_i = \frac{7.625 \text{ in.}}{12} = 0.6354 \text{ ft}$$

$$D_o = \frac{8.625 \text{ in.}}{12} = 0.7188 \text{ ft}$$

h_i in this case involves heat transfer by convection only, which has already been determined for identical conditions in Ex. 7-2 (page 142) as 133 Btu/(hr)(sq ft)(°F). h_o will involve both free convection and radiation and will therefore equal $h_c + h_r$. h_c may be calculated by the method explained in Art. 8-2, in the manner illustrated by Ex. 8-1. For the first approximation the mean temperature of the pipe surface and the main body of air will be $(600 + 70)/2 = 335°F$.

$$\text{At } 335°F, a_{\text{air}} = 0.25 \times 10^6 \qquad \text{(Fig. A-2)}$$
$$L = D_o = 0.7188 \text{ ft}$$
$$\Delta t = \text{approximately } 600 - 70 = 530°F$$
$$aL^3 \Delta t = 0.25 \times 10^6 \times 0.7188^3 \times 530 = 49,200,000$$

Since $aL^3 \Delta t$ is less than 10^9, Eq. (8-9) is applicable. C (by Table 8-2) = 0.45. k for air at 335°F (by Table A-2) = 0.0212.

$$h_c = C \frac{k}{L} (aL^3 \Delta t)^{1/4}$$

$$= 0.45 \frac{0.0212}{0.7188} (49,200,000)^{1/4}$$

$$= 0.01328 \times 83.75 = 1.11 \text{ Btu/(hr)(sq ft)(°F)}$$

h_r may be determined by Eq. (4-38) (page 74).

$$h_r = \frac{\sigma F_e F_A (T_s{}^4 - T_r{}^4)}{t_s - t_r}$$

where $\sigma = 0.173 \times 10^{-8}$.
$F_e = 0.79$ as illustrated in Ex. 4-1a (page 65).
$F_A = 1$.
$t_s = $ for first approximation, 600°F.
$t_r = 70°F$.
$T_s = $ for first approximation, $600 + 460 = 1060°$Fabs.
$T_r = 70 + 460 = 530°$Fabs.

$$h_r = \frac{0.173 \times 10^{-8} \times 0.79 \times 1 \times (1060^4 - 530^4)}{600 - 70}$$

$$= 3.05 \text{ Btu/(hr)(sq ft)(°F)}$$

$$h_o = 1.11 + 3.05 = 4.16 \text{ Btu/(hr)(sq ft)(°F)}$$

k for mild steel at 600°F (by Table 2-1, page 14) = 26.6

By substitution of the foregoing values into Eq. (11-5)

$$U_o = \frac{1}{\dfrac{0.7188}{0.6354 \times 133} + \dfrac{0.7188}{2 \times 26.6} \ln \dfrac{0.7188}{0.6354} + \dfrac{1}{4.16}}$$

$$= \frac{1}{0.0085 + 0.0135 \ln 1.131 + 0.2404}$$

$$= \frac{1}{0.0085 + 0.0017 + 0.2404}$$

$$= \frac{1}{0.2506} = 3.99 \text{ Btu/(hr)(sq ft)(°F)}$$

The need for a second approximation of the outside-surface temperature may be investigated as follows:

The quantity of heat passing through each of the three barriers is

$$U_o(600 - 70) = 3.99 \times 530 = 2115 \text{ Btu/(hr)(sq ft)}$$

In passing through the outside-surface resistance, this quantity 2115 Btu equals $h_o(t_s - t_r)$. Thus,

$$2115 = 4.16(t_s - 70) \qquad \text{or} \qquad t_s = \frac{2115}{4.16} + 70$$

$$t_s = 508 + 70 = 578°F$$

The only factor greatly affected by the change in surface temperature from 600 to 578° is h_r, which when calculated becomes 2.94 Btu/(hr)(sq ft)(°F). h_o then becomes 1.11 + 2.94 = 4.05. $1/h_o$ then becomes 1/4.05, or 0.2469, instead of 0.2404, and

$$U_o = \frac{1}{0.0085 + 0.0017 + 0.2469} = \frac{1}{0.2571}$$

$$= 3.89 \text{ Btu/(hr)(sq ft)(°F)}$$

By the second approximation it is seen that the over-all coefficient U_o is 2½ per cent lower than by the first approximation. A third approximation would reduce the value of U_o still further, but by an amount that would hardly justify the additional calculation required.

When it is desired to express U in terms of the inner surface, U_o, as found by Eq. (11-5), may be multiplied by D_o/D_i, or U_i may be expressed as

$$U_i = \frac{1}{\dfrac{1}{h_i} + \dfrac{D_i}{2k_m} \ln \dfrac{D_o}{D_i} + \dfrac{D_i}{D_o h_o}} \qquad \text{Btu/(hr)(sq ft)(°F)} \quad (11\text{-}6)$$

If the cylindrical wall is composed of successive layers of different material of appreciable thickness, as in the case of a heat-insulated pipe, the terms $(D_o/2k_m)$ ln (D_o/D_i) of Eq. (11-5) and $(D_i/2k_m)$ ln (D_o/D_i), which express the resistance to conduction, must be ropeated for each successive layer, with appropriate values of k and the inner and outer diameters D_i and D_o applied. Resistances due to thin layers, such as the films produced by the fouling of surfaces, are expressed directly by a numerical value which is the reciprocal of the conductance C.

Standards published by the Tubular Exchanger Manufacturers Association[4] for the design of heat exchangers call for an allowance to be made for the normal fouling of surfaces that is to be expected in operation. This allowance, expressed as the resistance of the film, ranges from 0.0005 for distilled water or for sea water at a temperature not exceeding 125°F to as high as 0.01 for dirty river water, such as in the Chicago Sanitary Canal, at low velocities and high temperatures. The introduction of an additional resistance of 0.01 into the denominator of Eq. (11-5) will be seen to reduce an over-all coefficient of 100 for a clean tube to a value of 50 for a fouled tube, whereas an over-all coefficient of 10 for a clean tube will be reduced by only 9 per cent by the same fouling resistance.

11-7. Mean Temperature Difference. Problems up to this point have dealt with heat transfer between regions at two different temperatures. Uniform temperatures have been assumed to exist throughout the extent of each region. In many cases of heat transfer, however, the temperatures in different parts of the hotter or of the cooler region are not the same. Along the length of a boiler tube, for example, the temperature of the water may be increasing while the temperature of the hot gases is decreasing. This will be true in *parallel flow*, i.e., when the water and hot gases are flowing in the same or parallel directions. If the water and hot gases travel in opposite directions (in *counterflow*), the hottest water and hottest gases will be at one end of the tube and the cooler water and cooler gases at the other end. A temperature gradient may be drawn to show the change in temperature from hot gas to water at any specific position along the tube, but the temperature gradients at different positions will not be alike.

Where either the hotter or the cooler medium is a saturated

vapor or a liquid at the boiling point corresponding to the existing pressure, its temperature will remain constant; but the temperature of the other medium will vary along the length of its travel.

Figures a, b, c, and d illustrate the changes in temperature that may occur in either or both mediums. The distances between the lines indicate the differences in temperatures of the

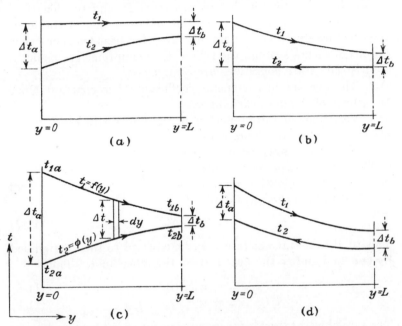

Fig. 11-5. Temperature differences in parallel and counterflow.

two mediums. The abscissa of each figure represents the length of travel. Figure a shows the hotter medium at a constant temperature; Figure b shows the cooler medium at a constant temperature. Figure c applies to parallel flow, and Figure d to counterflow. It should be observed that in each case the changing temperature is shown by a curved line, since it can be demonstrated by test and by mathematical analysis that the changing temperature is not a linear function of the travel along the length of the apparatus. It therefore follows that the mean temperature difference is not the arithmetic mean (i.e., not the average of the distances between the lines at the two ends). Instead, the true mean temperature difference, applying to all four parts of Fig. 11-5, a, b, c, and d, is the logarithmic mean temperature difference found as follows.

Assume two fluids separated by a solid boundary. The two fluids flow along the boundary with flow rates w_1 and w_2 in either parallel or counterflow. The temperatures of the two fluids are $t_1 = f(y)$ and $t_2 = \phi(y)$, as shown in figure c. The perimeter or width of the heat-exchange surface is P, and the length in the direction of flow is L. For the determination of the mean temperature difference commonly used for heat-transfer calculations in such a system, the following assumptions are made:

1. Steady-state conditions are established.

2. The over-all heat-transfer coefficient U is constant along the length of the heat exchanger.

3. Variations of the properties of the fluids with temperature are small enough to be neglected.

4. No phase change occurs in either fluid.

The mean temperature difference Δt_m to be determined is defined by the equation for the total rate of heat transfer from one fluid to the other, as follows:

$$q = UPL\,\Delta t_m \tag{11-7}$$

This total heat-transfer rate may be equated to the heat gained by one fluid and to the heat lost by the other fluid. Thus,

$$UPL\,\Delta t_m = w_1 c_{p1}(t_{1a} - t_{1b}) \tag{11-8}$$
$$UPL\,\Delta t_m = w_2 c_{p2}(t_{2b} - t_{2a}) \tag{11-9}$$

The rate of heat transfer across an element of the heat exchanger of length dy is

$$dq = UP\,\Delta t\,dy \tag{11-10}$$

This differential of the heat-transfer rate may be equated to the heat gained by the two fluids passing over this element and experiencing temperature changes dt_1 and dt_2, respectively. Hence

$$UP(t_1 - t_2)\,dy = -w_1 c_{p1}\,dt_1 \tag{11-11}$$
$$UP(t_1 - t_2)\,dy = w_2 c_{p2}\,dt_2 \tag{11-12}$$

Equations (11-8) and (11-9) may be combined to give

$$\frac{1}{w_1 c_{p1}} + \frac{1}{w_2 c_{p2}} = \frac{(t_{1a} - t_{2a}) - (t_{1b} - t_{2b})}{UPL\,\Delta t_m} \tag{11-13}$$

and Eqs. (11-11) and (11-12) may be combined to give

$$\frac{1}{w_1 c_{p1}} + \frac{1}{w_2 c_{p2}} = -\frac{d(t_1 - t_2)}{UP(t_1 - t_2)\,dy} \tag{11-14}$$

If Eqs. (11-13) and (11-14) are combined and the variables are separated, the result may be written in the integral form as follows:

$$-\int_{t_{1a}-t_{2a}}^{t_{1b}-t_{2b}} \frac{d(t_1 - t_2)}{t_1 - t_2} = \frac{(t_{1a} - t_{2a}) - (t_{1b} - t_{2b})}{\Delta t_m} \int_0^L \frac{dy}{L} \tag{11-15}$$

which produces the result

$$\Delta t_m = \frac{(t_{1a} - t_{2a}) - (t_{1b} - t_{2b})}{\ln \dfrac{t_{1a} - t_{2a}}{t_{1b} - t_{2b}}}$$

or
$$\Delta t_m = \frac{\Delta t_a - \Delta t_b}{\ln (\Delta t_a / \Delta t_b)} \tag{11-16}$$

where Δt_m = mean temperature difference, Δt_a = larger and Δt_b = smaller of the two temperature differences at the ends of the apparatus.

If the temperature difference Δt_a at one end of the apparatus is not more than 50 per cent greater than the temperature difference Δt_b at the other end, the arithmetic mean temperature difference will be within 1 per cent of the true, or logarithmic, mean temperature difference; but when the ratio $\Delta t_a / \Delta t_b$ is increased beyond 1.5, the error in the arithmetic mean as a measure of the temperature difference increases rapidly. The use of the arithmetic in place of the logarithmic mean temperature difference in simplified calculations is justifiable, therefore, only for small values of the ratio $\Delta t_a / \Delta t_b$.

In calculations of the mean temperature difference between the vapor and the cooling fluid in a condenser, the vapor is usually considered to be at the saturation temperature corresponding to the pressure, regardless of the amount of superheat that actually may be present. One of the reasons for this practice is that the superheat usually is removed before the vapor has come into contact with more than a small portion of the condenser surface. Furthermore, it has been found that where the desuperheating portion has been separated from the rest of the condenser so that its performance can be measured independently, the increase in heat transfer that might be expected in the desuperheating section (because of the increase in temperature difference due to superheat) is practically offset by reduction in the transmission coefficient for superheated as compared with saturated vapor.

11-8. Counterflow and Parallel Flow. The directions in which the fluids travel in a heat exchanger may have an important bearing upon the performance. Where there is a choice between the arrangements that provide for counterflow or for parallel flow of the two fluids, that arrangement which permits counterflow is usually preferred, for two reasons:

1. With counterflow the exchange of heat may raise the temperature of the cooler medium to more nearly the initial temperature of the hotter medium than is possible with parallel flow.

2. For exchange of the same amount of heat, a smaller extent of surface is needed in the case of counterflow than where the fluids travel in parallel directions.

The former advantage of the counterflow arrangement is readily observed by comparing c and d of Fig. 11-5. No matter how far the temperature lines of figure c (for parallel flow) are extended, the final temperature of the cooler medium can never reach the final temperature of the hotter medium; otherwise no exchange of heat could take place near the outlet end of the exchanger. If, on the other hand, the temperature lines of figure d (for counterflow) are extended, the final temperature of the cooler medium may even exceed the final temperature of the hotter medium, since a difference between the temperatures of the two fluids will exist at every point along the length of the exchanger.

The reason that less surface is required for the counterflow than for the parallel-flow arrangement for any certain heat exchange is not so apparent. The individual conductances and the over-all coefficient of heat transmission are in no way dependent upon the direction of flow. The mean temperature differences for the same initial and final temperatures, however, are not the same for parallel flow and for counterflow. Between the same temperature limits, the mean temperature difference for counterflow will always be greater than for parallel flow.

The heat exchange q is the product of the coefficient U and the surface area A and the mean temperature difference Δt_m, or in equation form $q = AU \Delta t_m$; but since, for any fixed value of q, U is the same for parallel flow as for counterflow, then A must vary inversely as Δt_m.` The extent of surface required for the parallel-flow arrangement is then greater than that required for counterflow in proportion to the inverse ratio of the mean temperature differences in the two cases.

The following example will show the significance of the direction of flow as it affects the extent of surface required.

Example 11-4. In a double-pipe exchanger, 500 lb of water per hour is to enter the outer circuit at 140° and leave at 100°F. In the inner circuit, 1,000 lb of water per hour will enter at 50° and leave at 70°. The heat exchange will equal $500(140 - 100)$, or $1000(70 - 50) = 20,000$ Btu/hr.

The temperature differences at the two ends, Δt_a and Δt_b, will be

For parallel flow:
$$\Delta t_a = 140 - 50 = 90°$$
$$\Delta t_b = 100 - 70 = 30°$$

For counterflow:
$$\Delta t_a = 140 - 70 = 70°$$
$$\Delta t_b = 100 - 50 = 50°$$

The logarithmic mean temperature differences,

$$\frac{\Delta t_a - \Delta t_b}{\ln (\Delta t_a/\Delta t_b)}$$

are

For parallel flow:
$$\frac{90 - 30}{\ln (90/30)} = \frac{60}{\ln 3} = \frac{60}{1.0986} = 54.6°F$$

For counterflow:
$$\frac{70 - 50}{\ln (70/50)} = \frac{20}{\ln 1.4} = \frac{20}{0.3365} = 59.4°F$$

The surface required for parallel flow, in inverse proportion to the mean temperature differences, will be 59.4/54.6, or 1.09 times the surface required for counterflow.

Further analysis would show that where the final temperatures of the fluids in the two circuits are brought close together, the advantage of the counterflow arrangement is made more pronounced.

11-9. Mean Temperature Difference in Multipass and Crossflow Heat Exchangers.

In many forms of heat exchangers the flow of the two fluids throughout the lengths of their circuits is neither continuously parallel flow nor continuously counterflow. The directions of flow of either or of both fluids may change during the travel through the exchanger. The heads of shell-and-tube heat exchangers are often provided with baffles, so arranged as to cause the fluid within the tubes to travel back and forth from one end of the exchanger to the other as many times as may be desired. In some cases longitudinal baffles within the shell likewise cause the fluid surrounding the tubes to travel the length of the shell a number of times. Multipass and crossflow designs are often preferred to a design embodying a strictly counterflow arrangement, for such reasons as a lower cost of manufacture, in spite of the greater extent of surface; the ease of removal of tube bundles for cleaning and for replacement; and the avoidance of excessive stresses due to uneven expansion of connected parts subjected to different temperatures.

Where the temperatures of both fluids are changing, it is apparent that the progressive changes will not be such as are shown by either figure c or d of Fig. 11-5. If one of the fluids is a boiling liquid or a condensing vapor, its temperature will remain substantially constant and the direction of flow of the second fluid will have no influence upon the mean temperature difference; but where the temperatures of both fluids are changing, the usual expression for the mean temperature difference [Eq. (11-16)] cannot be correctly applied to multipass exchangers without some modification.

Similarly, in the case of crossflow heat exchangers, in which the two fluids flow at right angles to each other, if the temperatures of both fluids are changing, Eq. (11-16) cannot be applied correctly without some modification.

Bowman, Mueller, and Nagle[5] have compiled the results found by numerous investigators who have studied the mean temperature differences for heat exchangers with neither parallel flow nor counterflow and have shown, by a series of graphs, the correction factors to be applied to Eq. (11-16) for the various flow arrangements. Three of these graphs are shown in Figs. 11-6, 11-7, and 11-8. Here, the correction factor F, which is read on the ordinates, is in the form of a multiplier to be applied to Eq. (11-16) calculated for *counterflow*. The values shown on the abscissas are for a dimensionless number P, equal to the ratio of temperature differences $(t_2 - t_1)/(T_1 - t_1)$; and the curves show a second dimensionless number R, equal to

$$\frac{T_1 - T_2}{t_2 - t_1}$$

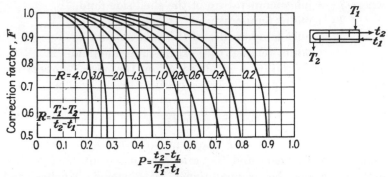

FIG. 11-6. Correction factor for heat exchangers with 1 shell pass and 2, 4, or any multiple of tube passes.

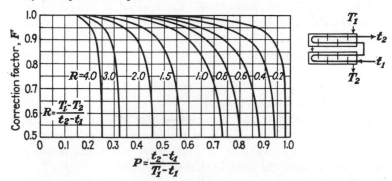

FIG. 11-7. Correction factor for heat exchangers with 2 shell passes and 4, 8, or any multiple of tube passes.

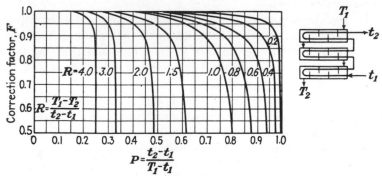

FIG. 11-8. Correction factor for heat exchangers with 3 shell passes and 6, 12, or any multiple of tube passes.

where T_1 = inlet temperature of the shell-side fluid, °F.

T_2 = outlet temperature of the shell-side fluid, °F.

t_1 = inlet temperature of the tube-side fluid, °F.

t_2 = outlet temperature of the tube-side fluid, °F.

Expressed in the form of an equation, the mean temperature difference in multipass and crossflow heat exchangers is

$$\Delta t_m = F \frac{(T_1 - t_2) - (T_2 - t_1)}{\ln \left[(T_1 - t_2)/(T_2 - t_1) \right]} \quad °F \qquad (11\text{-}17)$$

For additional values of F, applying to other arrangements of multipass exchangers and to various arrangements of crossflow exchangers, see the original paper by Bowman, Mueller, and Nagle.[5]

When Eqs. (11-16) and (11-17) are applied, it should be noted to what extent the heat exchanger may fail to conform with the requirements assumed in the derivations of these equations. In the derivation of Eq. (11-16) the following assumptions were made:

1. The over-all heat-transfer coefficient U must be constant throughout the heat exchanger.

2. The rate of flow and the specific heat of each fluid must be constant (except where heat is transferred by evaporation or condensation); i.e., each fluid must have the same phase condition throughout the heat exchanger.

3. The heat exchanger must be perfectly insulated.

4. The fluids in the two circuits must travel in the same or in opposite directions.

In the development of the correction factors for multipass exchangers it has further been assumed that there is the same amount of heat-transfer surface in each pass and that the temperature of the shell-side fluid in any shell-side pass is uniform over any cross-sectional area.

It may readily be observed that few of these requirements can be met with absolute accuracy, but the deviations from them are ordinarily too small to be of serious consequence in the calculation of the true mean temperature difference.

Problems

1. Calculate the over-all coefficient of heat transfer U for a piece of glass ⅛ in. thick, exposed to still air at 70°F on one side and 0°F air moving at 15 mph on the other side.

2. Find the amount of heat lost per hour through a square foot of building wall composed of 18-in. sandstone and $\frac{3}{4}$-in. metal lath and plaster separated by a $1\frac{1}{4}$-in. air space, when the temperature of the room is 70°F, the outdoors is 10°F, and the wind velocity is 15 mph.

3. Determine the percentage decrease in heat loss if the air space in the wall of Prob. 2 is filled with glass wool with a density of 1.5 lb_m/cu ft.

4. Calculate the heat loss per square foot of wall surface for a wall made of 1 in. of plaster and 9 in. of common brick. The air velocity over the surface of the brick is 15 mph; the outside air temperature is 40°F, and the surface temperature of the plaster facing the room is 70°F.

5. What is the rate of heat transfer in Btu/(hr)(sq ft) for a 6-in. concrete wall when the inside air temperature is 70°F and the outside air temperature is 10°F? The outside wind velocity is 15 mph.

6. Find the equivalent thickness of a cylinder of 1.00 in. i.d. and 2.00 in. o.d., as applied in a calculation of the conduction of heat based upon (a) the outer surface and (b) the inner surface.

7. A 6-in. polished copper sphere is heated electrically to maintain a surface temperature of 280°F. If the sphere is suspended in a large room where the air and wall temperatures are 80°F, how much heat, in Btu/hr, must be supplied to the sphere?

8. A cylinder 6 in. in diameter and 18 in. long is suspended horizontally in a large room. The air and wall surfaces of the room are at a temperature of 60°F while the surface temperature of the cylinder is 440°F. Compute (a) the surface coefficient due to free convection, (b) the heat transferred by free convection (neglecting the end areas), (c) the surface coefficient due to radiation if the surface emissivity is 0.75, and (d) the total heat transferred by free convection and radiation (neglecting the end areas).

9. A box made of copper plate with outside dimensions of 5 ft by 5 ft by 6 in. is suspended horizontally in a large room where the air and wall temperatures are 80°F. The box contains a heating coil which maintains a temperature of 500°F on all external surfaces. What is the total heat loss from the box in Btu/hr?

10. A room is to be maintained at 70°F by either a panel-heated floor with a surface temperature of 85°F or a panel-heated ceiling with a surface temperature of 120°F. If the coefficients of radiant-heat transfer for the floor and ceiling are, respectively, 0.90 and 1.0 Btu/(hr)(sq ft)(°F), what is the ratio of the heat supplied to the room by free convection and radiation from the floor to that supplied by the same means from the ceiling?

11. A furnace wall is made of 12-in. fire-clay refractory, burnt at 2426°F (a poor radiator). If the inside-surface temperature is 2000°F and the outside surface is exposed to still air at 80°F, determine (a) the rate of heat loss by convection and radiation from the furnace in Btu/(hr)(sq ft) and (b) the outside-surface temperature.

12. A cast-iron pipe which is 2 in. i.d. and 3 in. o.d. is used for heating water. The outside heating medium has a mean stream temperature of 300°F. It is found that the average stream temperature of the water inside the pipe is 100°F. The inside-surface coefficient has been calculated to be 300 Btu/(hr)(sq ft of inside surface)(°F), and the outside-surface coefficient is estimated to be 300 Btu/(hr)(sq ft of outside surface)(°F). Determine (a) the inside and outside surface temperatures of the pipe and (b) the amount of water in pounds per hour that can be heated from 80 to 120°F if the pipe is 15 ft long.

13. How much water can be heated per hour in a double-tube heat exchanger with parallel flow if superheated steam enters the inner tube at 480°F and leaves at 380°F with water entering the outer tube at 80°F and leaving at 280°F? The heat-exchange area is 10 sq ft, and the over-all coefficient from steam to water is 150 Btu/(hr)(sq ft)(°F).

14. A double-pipe heat exchanger, 20 ft long, arranged for counterflow, is made up of a copper tube (1.25 in. i.d. and 1.37 in. o.d.) placed within a well-insulated steel pipe (2.07 in. i.d.). Water enters the copper tube at 70°F and leaves at 80°F. The velocity through the tube is 3 fps. The annular space between the pipe and tube is filled with water which enters at 140°F and leaves at 100°F. Determine (a) the true mean temperature difference, (b) the heat transferred to the water in the tube, Btu/hr, (c) the over-all coefficient of heat transfer, Btu/(hr)(sq ft)(°F), based on the outside area of the tube, and (d) the over-all coefficient if the outer surface of the tube is fouled by a film which adds a resistance of 0.005.

15. Water at a temperature of 65°F enters the annular space of a double-tube counterflow-type heat exchanger and leaves at 109°F. If 400 lb of water per hour flows through the annular space and if the heat-transfer surface area is 10 sq ft what is the over-all coefficient of heat transfer, Btu/(hr)(sq ft)(°F)? Another fluid enters the inner tube at 135°F and leaves at 95°F.

16. Determine the length of a double-tube counterflow-type heat exchanger required to remove 30,400 Btu/hr from air flowing at 315 fps through the inner tube of steel which measures 1.6 in. i.d. and 1.8 in. o.d. The average stream temperature of the air is 740°F, and the average inside wall temperature of the tube is 160°F.

17. The temperatures on the outside surface at the two ends of a thin sheet-steel duct of large diameter are 90 and 80°F when the duct is exposed to air at 75°F. If the outside-surface coefficient is 0.80 Btu/(hr)(sq ft) (°F) and the resistance of the inside film is twice as great as that of the outside film, find (a) the amount of heat transferred per hour per square foot of surface and (b) the average temperature of the fluid flowing in the duct.

18. Carbon tetrachloride vapor at a temperature of 170°F is condensed in a mild-steel horizontal cylinder which is exposed to air at 70°F. The cylinder is 4 in. o.d. and 12 in. long and has a wall thickness of 1 in. Radiation from the outer surface is 1.05 Btu/(hr)(sq ft)(°F). Determine (a) the over-all coefficient U for the cylindrical wall based on the outer surface and (b) the temperatures of the inner and outer surfaces.

19. Determine the heat loss per linear foot of pipe for a 10-in. i.d. horizontal galvanized-steel duct 0.05 in. thick through which air is flowing at a velocity of 200 fpm and at an average temperature of 180°F when the temperature of the surrounding air is 70°F.

20. Repeat Prob. 19 but with the duct covered with asbestos paper $\frac{1}{32}$ in. thick.

21. Repeat Prob. 20 but with the asbestos paper painted with aluminum paint containing 26 per cent aluminum.

22. Water at an average stream temperature of 200°F flows at a rate of

2 fps through a 6-in. steel pipe 10 ft long which is painted with 26 per cent aluminum paint. If the pipe is located horizontally in a large room where the wall and air temperatures are 80°F, determine (a) the over-all coefficient of heat transfer based on the outer surface, (b) the temperature drop across the inside film, and (c) the temperature drop across the outside film.

NOTE: Six-inch standard pipe is 6.07 in. i.d. and 6.625 in. o.d.

23. Water enters a clean thin-wall horizontal copper tube of 1 in. i.d. at a temperature of 180°F and leaves at 140°F. The velocity in the tube is 4 fps. The tube is immersed in water maintained at an average temperature of 100°F, producing an outside-surface coefficient of 150 Btu/(hr) (sq ft)(°F). Determine (a) the true mean temperature difference, (b) the over-all coefficient of heat transfer, (c) the rate of heat transfer in Btu/ (hr)(sq ft of inside surface), and (d) the average temperature of the copper tube.

24. Water which flows through a pipe at a velocity of 2,140 fph is being heated from an entering temperature of 80°F to a leaving temperature of 180°F by saturated steam which condenses on the outside surface of the pipe. The inside cross-sectional area of the pipe is 0.0888 sq ft. The outside surface area of the pipe is 44.70 sq ft and the steam temperature is 300°F. Determine (a) the mean stream temperature, (b) the over-all coefficient of heat transfer, (c) the Reynolds number for flow in the pipe, and (d) the weight of steam condensed, pounds per hour, assuming that the entire amount of heat given up by the condensing steam is used to heat the water. The latent heat at 300°F is 910 Btu/lb.

25. The heat-transfer surface of a feed-water heater is made up of 1¼-in. horizontal steel pipes. The pipes are externally heated by saturated steam at a pressure of 134 psia (350°F). The water, with an initial temperature of 90°F and a final temperature of 190°F, flows through the pipes with a velocity of 4 fps. What length of pipe is required to produce this temperature rise?

NOTE: One and one-quarter-inch standard pipe is 1.38 in. i.d. and 1.66 in. o.d.

26. In a shell-and-tube heat exchanger, ammonia at a saturation pressure of 153 psia (80°F) is condensing on the outside of 1¼-in. horizontal steel pipes. The pipes are cooled by water flowing through them at an average temperature of 70°F and at a velocity of 1.0 fps. Determine the over-all coefficient of heat transfer based on the outside surface area if there are three pipes in each vertical row. Assume an outside surface temperature of 76°F.

NOTE: One and one-quarter-inch standard pipe is 1.38 in. i.d. and 1.66 in. o.d. The latent heat of vaporization of ammonia is 498.7 Btu/lb and the density of the liquid is 37.48 lb_m/cu ft.

27. A feed-water heater consists of 20 tubes, 1 in. i.d., arranged parallel to each other. Water enters the tubes at 60°F and leaves at 180°F, flowing at an average velocity of 1.25 fps. The tube bundle is surrounded by a shell in which saturated steam at 250°F is introduced. Assume the outside-surface coefficient based on the inside surface of the tube to be 1300 Btu/(hr)(sq ft)(°F). Determine the length required for each tube.

28. Heat is transferred from hot water to an oil in a double-pipe heat exchanger in which the fluids are in counterflow. The water enters the outer pipe at a temperature of 180°F and leaves at 114°F, whereas the oil enters the inner pipe at 70°F and leaves at 136°F. Calculate the true mean temperature difference.

29. A shell-and-tube-type water heater is to be designed to heat 4,000 lb of water per hour from 50 to 120°F when condensing steam at a temperature of 220°F. Values of h_i of 500 and h_o of 2000 Btu/(hr)(sq ft)(°F) are to be assumed, both based on the outside surface area of the tubes. Neglecting the effect of thermal resistance of the tubes, determine the tube surface area required.

30. In a counterflow heat exchanger oil enters at a temperature of 200°F and leaves at 110°F while water enters at 60°F and leaves at 80°F. The rate of water flowing through the exchanger is 1,000 lb/hr. Assuming the specific heat of oil to be 0.5 and water 1.0, compute (a) the ratio of oil flow to water flow, (b) the over-all coefficient of heat transfer if the area is 10 sq ft (c) the error involved if the arithmetic mean temperature were used, and (d) the true mean temperature difference for parallel flow.

31. What will be the mean temperature difference to be used for design purposes if the process described in Prob. 28 takes place in a shell-and-tube-type heat exchanger in which the oil flows through the tubes in eight passes and the surrounding water flows through two shell passes?

32. Heat is transferred at the rate of 300,000 Btu/hr through 100 sq ft of surface in a shell-and-tube-type heat exchanger in which oil flows through the tubes in four passes and water flows through the shell in two passes. The oil enters at 80°F and leaves at 120°F, while the water enters at 150°F and leaves at 100°F. Determine (a) the temperature difference to be used for design purposes and (b) the over-all coefficient U for the heat-transfer surface.

33. A hollow copper cylinder having an inside diameter of 3.0 in., a wall thickness of 0.5 in., and a length of 5.0 in. is used for the combustion chamber of a small experimental rocket motor. The motor is cooled by water flowing through an annular passage surrounding the combustion chamber. The radial width of the annular passage is 0.15 in. It is determined experimentally that for a water rate of 30 gpm and an inlet water temperature of 50°F the exit water temperature is 65°F and the average outside surface temperature of the combustion chamber is 400°F. Determine (a) the average inside surface temperature of the combustion chamber and (b) the outside-film coefficient. (c) What is the outside-film coefficient, based on Eq. (7-5), and (d) if the average temperature of the gases in the combustion chamber is 6000°F, what is the value of the inside-film coefficient?

REFERENCES

1. Rowley, F. B., and A. B. Algren: Thermal Resistance of Air Spaces, *Trans. ASHVE*, **35**, 165–181 (1929).
2. Standard Test Code for Heat Transmission through Walls, *Trans. ASHVE*, **34**, 253–268 (1928).

3. Nicholls, P.: Measuring Heat Transmission in Building Structures and a Heat Transmission Meter, *Trans. ASHVE*, **30**, 65–104 (1924).
4. Standards of Tubular Exchanger Manufacturers Association, New York, 1949.
5. Bowman, R. A., A. C. Mueller, and W. M. Nagle: Mean Temperature Difference in Design, *Trans. ASME*, **62**, 283–294 (1940).

CHAPTER 12

APPLICATIONS OF THE PRINCIPLES OF HEAT TRANSFER TO DESIGN PROBLEMS

12-1. General Method. The foregoing chapters have dealt with the fundamental principles that generally apply to problems of heat transfer. When these principles are applied to problems of design, the accuracy of results may at times be questionable because of the use of insufficient or inaccurate data relating to the physical properties of matter. Again, such uncertain influences as the cleanliness and roughness of surfaces and their tendency to become wetted and variations in their shapes and positions may not be properly evaluated; yet the result may be of considerable value, particularly in the development of new equipment.

In the design of many forms of equipment the first attempt to apply the principles of heat transfer has produced results that have been found to be far from correct. Tests of the equipment, however, have shown wherein the inaccuracies lay and have produced the necessary data whereby later designs have been worked out accurately. By the correlation of performance data with design data, much information has been accumulated relating to the design of certain forms of equipment, such as feed-water heaters, condensers, boilers, internal-combustion-engine cylinders, fan coils, and electrical transformers, motors, and generators.

In this chapter a few examples are presented that illustrate the manner and extent to which the principles of heat transfer may be applied in the design of certain forms of heat exchangers. The purposes in presenting these examples are both to give the student further practice in applying the principles with which he is already familiar and to illustrate the way in which he may find it necessary to modify or build upon those principles when applying them to design problems.

Many practical problems in heat transfer can be solved by making some original assumptions based upon performance data and then later modifying these assumptions to comply with the principles of heat transmission.   A thorough knowledge of the principles makes it possible to arrive at reasonable assumptions for the first approximation.   It should be remembered that any theoretical design is usually modified to comply with manufacturing standards in order to make use of standard tube diameters, standard tube lengths, standard plate thicknesses, etc. Furthermore, a design may have to be altered in order that an economical heat exchanger may be produced.   Such matters as the cost of pumping fluids and the initial investment must be balanced.

CALCULATION OF HEAT TRANSFER IN A FEED-WATER HEATER

The following example has been selected to illustrate the calculations involved in the selection of a closed feed-water heater.   In this example little attention is paid to modifications required to comply with manufacturing standards, and no attempt is made to obtain an economic balance between the cost of operation and the initial investment.   The solution of the problem is confined primarily to the heat-transfer principles involved.

Example 12-1.   It is desired to select a closed-type feed-water heater to serve a 100-hp boiler.   The boiler operates at 125 psig and delivers steam containing 1 per cent moisture.   The heater is to be capable of heating enough water to permit the boiler to operate at 200 per cent of its rating. The water is to be heated from 50 to 180°F, and the necessary heat is to be furnished by saturated steam at 230°F.   Because of space limitations a vertical-type heater not over 10 ft high must be selected.

It is first necessary to determine by the use of steam tables the amount of heat that is to be transferred.   The enthalpy of the feed water entering the boiler is approximately $180 - 32 = 148$ Btu/lb.   The enthalpy of the steam leaving the boiler is 1184 Btu/lb.   The heat added to the steam in the boiler is, therefore, $1184 - 148 = 1036$ Btu/lb.   Then the weight of steam to be evaporated at 200 per cent of rating is

$$\frac{200 \times 33,470}{1036} = 6461 \text{ lb/hr}$$

where 33,470 Btu is the heat equivalent of a boiler horsepower hour.
The total rate of heat transfer in the feed-water heater is

$$6461(180 - 50) = 839,930 \text{ Btu/hr}$$

The total surface area required in the heater may be calculated by the equation

$$q = U_o A \, \Delta t_m \quad \text{or} \quad A = \frac{q}{U_o \, \Delta t_m} \tag{12-1}$$

where q = total rate of heat transfer, Btu/hr.

U_o = over-all coefficient, Btu/(hr)(sq ft of outside tube surface)(°F of mean temperature difference).

A = outside tube surface, sq ft.

Δt_m = logarithmic mean temperature difference between the steam and the water.

In this case q is 839,930 Btu/hr and, by Eq. (11-16),

$$\Delta t_m = \frac{(230 - 50) - (230 - 180)}{\ln \left[(230 - 50)/(230 - 180) \right]} = 101.5°F$$

The value of U_o cannot be determined until the velocity of the water in the tubes is known, but the water velocity depends upon the tube size and number of tubes. Two methods of attack may be followed: Either a velocity and a tube size may be selected, from which the tube length may be calculated, or a tube length and size may be selected, from which the number of tubes may be calculated. In either case some preliminary assumptions will have to be made to obtain the first approximation. Then by a process of successive approximations, the combination of tube area and velocity required to give the desired temperature rise and total rate of heat transfer may be calculated.

In this case let it be assumed that 1-in. 14 Birmingham Wire Gauge brass tubes will be used. Since the total length of the heater must not exceed 10 ft, a tube length of 6 ft may be assumed for the first approximation. From Table 11-2 it is seen that the value of U_o for feed-water heaters will likely lie between 200 and 1500. Owing to the large temperature rise and the limited tube length, a low water velocity will have to be used and therefore a low value of U_o may be expected. If a value of 200 is chosen for the first rough calculation, the required tube area may be calculated from Eq. (12-1). Thus,

$$A = \frac{839,930}{200 \times 101.5}$$
$$= 41.4 \text{ sq ft}$$

The dimensions of a 1-in. 14 BWG brass tube are 1 in. = 0.0833 ft o.d., 0.834 in. = 0.0695 ft i.d., 0.00379 sq ft cross-sectional area, and 0.2618 sq ft of outside surface per foot of length. The number of tubes required is $41.4/(0.2618 \times 6) = 26.4$, say 26. If a two-pass* heat exchanger is used with 13 tubes in each pass, then the water velocity will be

$$\frac{6,461}{61.54 \times 0.00379 \times 13} = 2,130 \text{ fph} \quad \text{or} \quad 0.592 \text{ fps}$$

* This results in a velocity that is two times the velocity in a single-pass exchanger.

This is well above the critical velocity for a 1-in. tube, and therefore Eq. (7-14) may be used to calculate the value of h_i, the inside-film coefficient. This is

$$h_i = 0.00134(t + 100) \frac{V^{0.8}}{D^{0.2}} \qquad (12\text{-}2)$$

The mean temperature of the main body of the water in the tubes is obtained by subtracting the logarithmic mean temperature difference from the steam temperature; its value is $230 - 101.5 = 128.5°F$. Since the resistance of the inside water film may be estimated to be of the order of four times the resistance of the outside film, then the tube-wall temperature will be approximately equal to the steam temperature minus one-fifth of the total temperature drop. This is $230 - (101.5/5) = 209.7°F$. The average temperature of the inside film, which is the average of the wall temperature and the temperature of the main body of water in the tubes, is equal to

$$\frac{209.7 + 128.5}{2} = 169.1°F$$

Then

$$h_i = 0.00134(169.1 + 100) \frac{2{,}130^{0.8}}{0.0695^{0.2}}$$
$$= 282 \text{ Btu/(hr)(sq ft)(°F)}$$

The value of h_o, the outside-film coefficient, is found by Eq. (10-4), which, when applied to a vertical tube, is

$$h_o = 134.8 \left(\frac{k^3 \rho^2 r}{L \mu \, \Delta t} \right)^{1/4} \qquad (12\text{-}3)$$

Since the average temperature of the outside film is

$$\frac{230 + 209.7}{2} = 219.9°F$$

then the values of the terms to be substituted in the equation for h_o are

$$
\begin{array}{ll}
k = 0.395 & \mu = 0.65 \\
\rho = 59.63 & L = 6 \\
r = 958.8 & \Delta t = 230 - 209.7 = 20.3
\end{array}
$$

and therefore

$$h_o = 134.8 \left(\frac{0.395^3 \times 59.62^2 \times 958.8}{6 \times 0.65 \times 20.3} \right)^{1/4}$$
$$= 970 \text{ Btu/(hr)(sq ft)(°F)}$$

To obtain a more accurate value of U_o it is necessary to use Eq. (11-5), which is

$$U_o = \frac{1}{\dfrac{D_o}{D_i h_i} + \dfrac{D_o \ln (D_o/D_i)}{2k_m} + \dfrac{1}{h_o}} \qquad (12\text{-}4)$$

where

$$\frac{D_o}{D_i h_i} = \frac{0.0833}{0.0695 \times 282} = 0.00425$$

$$\frac{D_o \ln (D_o/D_i)}{2k_m} = \frac{0.0833}{2 \times 69} \ln \frac{0.0833}{0.0695} = 0.00011$$

$$\frac{1}{h_o} = \frac{1}{970} = 0.00103$$

and therefore

$$U_o = \frac{1}{0.00425 + 0.00011 + 0.00103}$$

$$= \frac{1}{0.00539} = 185.5 \text{ Btu}/(\text{hr})(\text{sq ft})(°F)$$

This value is somewhat lower than the value originally assumed; hence, it is necessary to recalculate the area required. By Eq. (12-1), the new area required is

$$A = \frac{839,930}{185.5 \times 101.5} = 44.6 \text{ sq ft}$$

This larger area may be obtained by increasing the number of tubes or by increasing their length. If the number of tubes is increased, the velocity will be decreased and consequently the value of U_o will decrease.

At this point the practical designer would study the costs* involved, to determine which course to follow. The probable procedure would be to study the total costs for several velocities, and the velocity showing the lowest total cost would be selected. In this case let it be assumed that the larger area will be obtained by increasing the tube length. The required tube length will be increased in the ratio of the new area to the area originally assumed. Thus the tube length will be $6 \times (44.6/41.4) = 6.45$ ft, say, $6\frac{1}{2}$ ft.

For a first approximation, therefore, the heater to be selected will be one having two passes, each pass containing thirteen 1-in. brass tubes $6\frac{1}{2}$ ft long. This will fit into the available space and will provide the desired temperature rise. The use of $6\frac{1}{2}$-ft tubes instead of 6-ft tubes will result in a slightly lower value of h_o. The value of h_o used in Eq. (12-4) should be decreased by the ratio $(6/6.5)^{\frac{1}{4}}$. The value of this fraction is 0.98, or, in other words, h_o should be reduced by 2 per cent. This change would have practically no effect on the value of U_o. A larger effect may be caused by the ratio of film resistances originally assumed in calculating the wall temperature. From the separate resistances calculated in Eq. (12-4) it is seen that the temperature drops through the various resistances will be

a. Drop through outside film $= 101.5 \left(\dfrac{0.00103}{0.00539}\right) = 19.4°F$

b. Drop through tube wall $= 101.5 \left(\dfrac{0.00011}{0.00539}\right) = 2.1°F$

c. Drop through inside film $= 101.5 \left(\dfrac{0.00425}{0.00539}\right) = 80.0°F$

---

* For studies of economic balance in heat-transfer apparatus see Refs. 1 and 2.

Thus the actual outside-wall temperature is $230 - 19.4 = 210.6°F$, and the average temperature of the outside film will be $(230 + 210.6)/2 = 220.3°F$ instead of $219.9°F$, which was initially assumed. This difference will not appreciably affect the value of h_o previously determined, since the properties of water are only slightly affected by small temperature changes.

The inside-wall temperature is $210.6 - 2.1 = 208.5°F$, and the average inside-film temperature is $(128.5 + 208.5)/2 = 168.5°F$ instead of $169.1°F$, which was used in the first approximation. The difference in the value of h_i originally calculated and the value that would be obtained by a second approximation is so small that the value of U_o is not going to be materially changed.

In this case nothing would be gained by another approximation, since the resulting tube length and tube number would be the same as those used in the first approximation, viz., 26 tubes each $6\frac{1}{2}$ ft long. If, however, the wall temperature had been substantially different from that originally assumed, then it would have been necessary to recalculate the value of U_o and perhaps change the tube length.

CALCULATION OF HEAT TRANSFER IN A SURFACE CONDENSER

12-2. General. A surface condenser, such as is used in steam power service (Fig. 12-1), is essentially a bank of water tubes

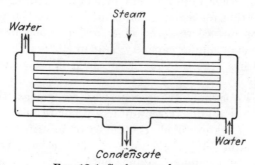

FIG. 12-1. Surface condenser.

enclosed within a steam space formed by a metal shell. Water enters the space between the head of the shell and the adjacent tube sheet and flows through the tubes to the opposite end of the condenser in a single pass. Otherwise, the water space may be so baffled as to cause the water to traverse the length of the condenser two or more times. Most surface condensers are of either the single-pass or the two-pass arrangement and are made with the tubes in a horizontal position. The water outlet is always at a higher elevation than the inlet so as to make sure that the tubes will be at all times filled with water.

The steam enters the shell at the top and flows downward over and between the tubes, the condensate gravitating to the hot well and condensate pump below. The single-pass condenser is usually applied where a liberal supply of cooling water is available and where the pumping head and cost of power are low, whereas the two-pass arrangement is more desirable where the opposite conditions prevail.

12-3. Factors Affecting Heat Transfer. It is apparent from the study of convection inside tubes and condensation on outer surfaces that the rate of heat transfer is dependent upon the velocity of the fluid in the tubes and upon the shape and condition of the surfaces, in addition to the difference in the temperatures of the two fluids. The necessity for crowding a large number of tubes in a limited space in a steam condenser introduces a number of additional factors, chief of which are the problems of effective distribution of the steam to all surfaces and the rapid removal of air which, if not removed, would form an insulating film on tube surfaces. Observance of these factors has brought about a progressive improvement in condenser design in recent years, accounting for a reduction of not far from 50 per cent in the extent of surface required in a modern condenser as compared with one of similar capacity built in 1920. The individual factors affecting heat transfer in steam condensers will be discussed briefly.

a. Water Velocities. It is generally recognized that water velocities of less than 3 ft/sec are likely not to provide uniform distribution of water to all the tubes of a surface condenser. Above this minimum, the choice of the velocity is governed primarily by the economic balance between the increase in heat transfer (and the resulting decrease in initial cost) with increase in velocity and the offsetting increase in the cost of pumping.

Where the water tends to deposit foreign matter in the tubes, a high velocity is desirable, but under many conditions increased corrosion of the tubes places a definite limit on the maximum velocity. With corrosive water or salt water, many engineers consider that a velocity of 7 ft/sec should not be exceeded regardless of the pumping costs. Ordinarily with good water a velocity of from 6 to 7 ft/sec is considered good practice for a two-pass condenser, and a velocity of from 7 to 7½ ft/sec is found suitable for a single-pass condenser.

  b. Cleanliness of Tubes.   Condenser ratings are generally based upon heat-transfer values for commercially clean tubes. Tests on originally clean condenser tubes, however, show a progressive increase in the thermal resistance of the film on the inside surface of the tubes as the tests proceed, even though the cooling water is reasonably free from foreign matter. A thin coating of sediment deposited by the water, and in some cases iron oxide from the piping system, gradually adheres to the surface. Even where condensers in service are to be cleaned with reasonable frequency, an allowance in capacity must be made for the decrease in heat transfer caused by the deposit on the water side. Good practice sets the reduction in the over-all coefficient of heat transfer due to fouling of the water-side surface of the tubes at 15 or 20 per cent, and more if poor maintenance or exceptionally dirty water is expected. A more logical procedure in making this allowance would be to apply a suitable corrective factor to the inside-surface coefficient only. This factor for fairly clean water, as judged from analyses of test results, is of the order of 80 per cent if the surface coefficient for clean tubes is calculated by McAdams' equation [Eq. (7-5)], or 70 per cent if the coefficient is calculated by von Kármán's equation [Eq. (7-8)]. Otherwise, allowance for fouling of the water-side surface may be made by applying a fouling factor in the form of an additional resistance in the calculation of the over-all coefficient.[3]

  When condensers are connected to the exhaust of reciprocating engines, lubricating oil entrained in the steam is soon deposited on the outside of the tubes. The coating of oil found in this type of service causes a further decrease in capacity, usually considered to be 35 per cent. Here again, a more logical procedure is to apply an additional fouling factor for representing the resistance of the oil film.

  c. Steam Distribution and Air Removal.   The volume of steam entering a condenser is ordinarily so great as to make a high velocity in the vicinity of the steam inlet inevitable, regardless of the size of the inlet. High steam velocity, although conducive to high rates of heat transfer and rapid removal of entrained air from heat-transfer surfaces, makes the problem of steam distribution difficult. This is true, especially since any baffles that may be used for directing the flow of steam are undesirable from the standpoint of the pressure loss that they create and because of space limitations. The presence of a cold tube surface in a steam space offers no assurance that steam will come into intimate contact with the surface at a rate sufficient to produce effective heat transfer. The quantity of entrained air, although ordinarily small in comparison with the amount of steam entering the condenser, may be large in comparison with the amount of steam reaching the lower tubes and is usually sufficient to produce an insulating air film on any surface over which the steam fails to flow at a brisk rate. A brisk downward flow of steam is of advantage further in effecting rapid removal of condensate from the tubes or at least a thinning of the film of condensate that accumulates on the tube surface. It is undoubtedly owing to the effects of steam velocity that tests of condensers generally show values for the over-all coefficient of heat transmission that increase with increase in steam loading.

Various schemes have been adopted by manufacturers of condensers for assuring effective distribution of steam and removal of air.[4] One is by the omission of certain rows of tubes from the upper portion of the tube bank in order to provide lanes through which steam may more readily reach the lower tubes. Another is by making the cross sections of the shell and the tube bank more nearly V- or heart-shaped than circular. The heart shape, because of its decreasing width from top to bottom, provides a decreasing area for steam flow as the volume of steam is progressively reduced by condensation. The reduction in flow area may be further accomplished by a decrease in tube spacing from wide centers in the top rows to close centers in the bottom.

Reference to trade literature will show other improvements in condenser design that aim to reduce the cost of vacuum-pump operation; but since these improvements are not related directly to the problem of heat transfer, they will not be discussed here.

12-4. Heat-transfer Coefficients. Before it became common practice to break down the over-all coefficient of heat transmission into separate conductances, the over-all coefficient for a number of designs of surface condensers was observed to vary approximately as the square root of the velocity of the water in the tubes.[5] This observation led to general adoption of the "square-root" rule, a rule that is still commonly applied in the rating of condensers.* In the light of present-day knowledge of surface resistances, however, it is not surprising that in many individual tests the variation in the over-all coefficient has failed to follow the square-root rule, since any change in design, such as an improvement in steam distribution, that affects one and only one of the thermal resistances will necessarily alter the relationship between the over-all coefficient and the water velocity. The more logical procedure, therefore, in any study of heat transfer in surface condensers is to observe the separate resistances to the flow of heat from the steam to the water.

12-5. Thermal Resistances. Of the three thermal resistances, viz., the inside film, the tube itself, and the outside film, the resistance of the inside film is ordinarily the greatest, at least in the case of the upper tubes. The resistance of the thicker film of condensate and air on lower tubes, however, may very likely exceed the resistance of the inside film.

a. Inside-surface Coefficient. The inside-surface coefficient may be calculated, for clean tubes, with reasonable accuracy by applying any of the expressions in Art. 7-6. By reference to Fig. 7-1 it may be noted that for the usual range of cooling-water temperatures (and corresponding Prandtl numbers) the lowest or most conservative values will be found by use of McAdams' and Prandtl's equations [Eqs. (7-5) and (7-7)] whereas von Kármán's equation [Eq. (7-8)] will yield a result that is approximately 10 per cent greater. Tests made by Ferguson and Oakden[7] on single tubes of various sizes, electrically heated and cleaned prior to each test, show surface coefficients in most cases even slightly greater than the values found by von Kármán's equation. The choice of the equation, therefore, may be made according to whether a conservative figure or the maximum value substantiated by tests is desired.

---

* Standard limits for the ratings of condensers have been set up by the Condenser Section of the Heat Exchange Institute. By these limits, guarantees will not be made for operation with a difference of less than 5° in the temperatures of the steam and the outlet cooling water and an oxygen content of the condensate of less than 0.03 cu cm/liter. Over-all heat-transfer coefficients for water at 70°F inlet temperature, flowing through commercially clean tubes of various diameters, are established. These coefficients are assumed to vary as the square root of the water velocity. Correction curves are applied for variations in inlet water temperature and condenser loading. See Ref. 6.

b. Resistance of the Tube.   The thermal resistance of the tube itself is small but hardly negligible.   Condenser tubes are commonly made of brass (70 per cent copper, 30 per cent zinc) or of Muntz metal (60 per cent copper, 40 per cent zinc).   Where the cooling water is contaminated by sewage or other foreign matter or where sea water is used, the tubes are commonly made of admiralty metal (70 per cent copper, 29 per cent zinc, 1 per cent tin).   For all these alloys at usual temperatures the thermal conductivity k is approximately 65.   Tubes are generally $3/4$, $7/8$, or 1 in. o.d. and 0.049 in. (18 BWG) thick, although $5/8$ in. o.d. tubes are sometimes used in small condensers and thicker tubes are used where rapid deterioration is anticipated.   The choice of the tube size is usually made according to the fouling characteristics of the water, the larger sizes being preferred where the tendency to foul is greater.

c. Outside-surface Coefficients.   The greatest uncertainty in the calculation of heat transfer in condensers lies in the prediction of the value of the surface coefficient for the steam side of the tubes.   Because of the high and variable velocity of the steam in the condenser, no formula for the surface coefficient, such as Eq. (10-2) discussed in Art. 10-2, can be applied here.   Tests by Ferguson and Oakden[7] on a vertical row of 6 condenser tubes have shown higher values of the surface coefficient for steam at 90°F than at 120°F when the same weight of steam was being supplied in each test.   This is at variance with the theory upon which the equations of Chap. 10 are based and can be explained only by the observation that at the lower temperature the increase in velocity due to increase in specific volume of the steam more than offsets the effect of the increased viscosity of the film.   Ferguson's and Oakden's tests, although valuable in showing the manner in which the surface coefficient is influenced by such factors as the steam velocity, the rate of heat transfer, the condition of the surface, the position of the tube, and the percentage of air present, naturally fail to develop for the surface coefficient any average value that can be applied in the design of a condenser.   The nearest approach to such a value was made in tests of a specially designed condenser built with 18 tubes in staggered arrangement, seven rows deep and alternately two and three rows wide.   Here, average values of the order of 2500 to 2900 Btu/(hr)(sq ft of outer surface)(°F) were obtained; but since only a part of the steam entering the condenser was condensed, and since the surface coefficient for individual rows of tubes decreased from the top to the bottom row, the values can be regarded as representative only for the upper rows of tubes in a condenser of commercial form.

Contrasted with Ferguson's and Oakden's values it may be observed that for a tube completely blanketed with a film of air, as is conceivable in the case of a tube near the bottom of a condenser, the surface coefficient would be of the order of only 1 or 2 Btu instead of 2500 to 2900.

In view of the lack of a satisfactory value for the outside-surface coefficient determined directly by tests, it would appear that a suitable value for any commercial form of condenser can best be obtained from the over-all coefficient established by tests of a condenser of similar design. Thus, if the value of U_o of Eq. (11-5)

$$U_o = \cfrac{1}{\cfrac{D_o}{D_i h_i} + \cfrac{D_o}{2k_m} \ln \cfrac{D_o}{D_i} + \cfrac{1}{h_o}}$$

is established by test, the value of h_o for the same rate of condensation may be determined by substitution of the values of U_o, h_i, k_m, D_i, and D_o into the equation.

U_o is ordinarily found from test data by dividing the heat in Btu/hr removed from the steam or added to the water per square foot of outer surface by the logarithmic mean temperature difference between the saturation temperature of the steam at condenser inlet pressure and the temperature of the cooling water as it enters and leaves the condenser. At low rates of condensation all the steam may be condensed before reaching the lowest tubes, and the temperature of the condensate in the hot well may be considerably below the steam temperature, whereas at higher rates of condensation the conversion of the velocity pressure of the steam into static pressure has been shown in some instances to cause the temperature of the condensate to rise several degrees above the steam temperature.[8]   In spite of these differences in steam and hot-well temperatures it is not surprising that the most consistent results are found when the hot-well temperature is disregarded in the calculation of the mean temperature difference, for it is unlikely that hot-well temperatures can extend beyond a very small portion of the condenser surface.

As already noted, a proper value of h_i for substitution into Eq. (11-5) can be determined by applying Eq. (7-5) together with an appropriate cleanliness factor.

Application of this method by the authors of this text to data obtained in an extensive series of tests of a 50,000-sq ft condenser, reported by G. H. Van Hengel,[8] yields consistent values of h_o that are found to vary with the rate of condensation but that appear to be substantially independent of all other variable factors such as steam and water temperatures, the mean temperature difference, and the water velocity.

At a load of 8 lb of condensate per hour per square foot of outside tube surface the value of h_o derived from these tests is approximately 1000 Btu/(hr)(sq ft)(°F).* This value is based upon an assumed cleanliness factor of 90 per cent applied to the inside-surface coefficient as calculated by Eq. (7-5). If a lower value is assumed for the cleanliness factor, a higher value of h_o results but the calculated values of the over-all coefficient U show poorer agreement with the corresponding test values. In this series of tests, which extended over a 10-month period, chlorinated cooling water was used. Accordingly, little variation in the cleanliness of the tubes would be expected. The percentage of variation in the value of h_o found for other loadings is shown in Fig. 12-2.†

Application of the foregoing method to the 31 tests for which Van Hengel has published test data has yielded calculated values of the over-all coefficient U that do not differ from the test values of U by more than 10 per cent in 28 of the 31 tests. No reason has been found for the apparently inconsistent values derived from the remaining 3 tests.

* The value of 1000 Btu/(hr)(sq ft)(°F) for the outside-surface coefficient, when it is applied to Eq. (11-5), yields values for the over-all coefficient that are lower than the limiting values shown by the "square-root" curve adopted by the Condenser Section of the Heat Exchange Institute. This is true even when the values assigned to the inside-surface coefficient are the maximum values substantiated by tests.

† Although Fig. 12-2 has been derived from a series of tests of only one condenser, it is probably fairly representative of the manner in which the value of h_o changes with the loading in other condensers of usual design.

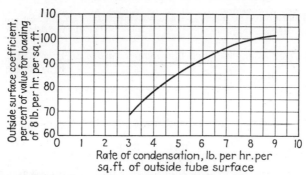

Fɪɢ. 12-2. Variation in outside-surface coefficient with rate of condensation.

The following example will illustrate the application of the foregoing principles in the design of a condenser.

Example 12-2. A condenser is to be designed to serve a turbogenerator unit that will exhaust 33,000 lb of steam per hour. A vacuum of 28.0 in. Hg (referred to a 30-in. barometer) is desired. An adequate but not liberal supply of slightly muddy cooling water at a maximum temperature of 80°F is available. A two-pass condenser of normal design is contemplated.

Data obtained in the operation of a condenser of a design similar to that which is contemplated show the rate of heat transfer to be 6855 Btu/ (hr)(sq ft of outer surface) under the following conditions:

Condition of tubes—clean.
Diameter of tubes—0.75 in. o.d., 0.652 in. i.d.
Velocity of water = 5.10 fps.
Condenser pressure—1.66 in. Hg.
Corresponding steam temperature—95.0°F.
Temperature of inlet cooling water—73.3°F.
Temperature of outlet cooling water—87.0°F.

Nᴏᴛᴇ: Where no determination of the quality of the exhaust steam is made, the heat removed by the condenser is assumed to be 950 Btu/lb of steam exhausted from turbines and 1000 Btu/lb for engines. On this basis the weight of condensate leaving the condenser, corresponding to 6855 Btu/(hr)(sq ft), will be $6855/950$ = 7.22 lb/(hr)(sq ft).

Solution. The solution will involve the following steps:

1. Calculation of the over-all coefficient U_o for test conditions. The mean temperature difference, for test conditions, is

$$\frac{(95.0 - 73.3) - (95.0 - 87.0)}{\ln\,[(95.0 - 73.3)/(95.0 - 87.0)]} = 13.75°\text{F}$$

$$U_o = \frac{6,885}{13.75} = 498.5 \text{ Btu/(hr)(sq ft)(°F)}$$

and the over-all thermal resistance,

$$\frac{1}{U_o} = 0.002006 \text{ (hr)(sq ft)(°F)/Btu}$$

2. Calculation of the inside-surface resistance $D_o/D_i h_i$ for test conditions with clean tubes.

h_i[by Eq. (7-5)] at an estimated film temperature of 85°F = 1181 Btu/(hr)(sq ft)(°F)

$$\frac{D_o}{D_i h_i} = \frac{0.75}{0.652 \times 1181} = 0.000975 \ (\text{hr})(\text{sq ft})(°F)/\text{Btu}$$

3. The thermal resistance of the tube,

$$\frac{D_o}{2k_m} \ln \frac{D_o}{D_i} \text{ [from Eq. (11-5)]} = \frac{0.75}{2 \times 65} \ln \frac{0.75}{0.652} = 0.000067$$
$$(\text{hr})(\text{sq ft})(°F)/\text{Btu}$$

4. The outside-surface resistance, $1/h_o$, is established by substituting into Eq. (11-5) the values found in steps 1, 2, and 3. Equation (11-5), for convenience, may be written in the form

$$\frac{1}{U_o} = \frac{D_o}{D_i h_i} + \frac{D_o}{2k_m} \ln \frac{D_o}{D_i} + \frac{1}{h_o}$$

Thus

$$0.002006 = 0.000975 + 0.000067 + \frac{1}{h_o}$$

and

$$\frac{1}{h_o} = 0.000964 \ (\text{hr})(\text{sq ft})(°F)/\text{Btu}$$
$$h_o = 1037 \ \text{Btu}/(\text{hr})(\text{sq ft})(°F)$$

5. $D_o/D_i h_i$ may now be calculated for the specified operating conditions. Condensers of the two-pass type with from 500 to 5,000 sq ft of surface are ordinarily fitted with ¾-in. tubes; from 5,000 to 10,000 sq ft, with ⅞-in. tubes; and larger sizes almost universally use 1-in. tubes. For rough calculations 1 sq ft of surface may be assumed to condense 8 lb of steam per hour. Accordingly, to condense 33,000 lb of steam per hour, the required surface will be of the order of 4,120 sq ft indicating the use of ¾-in. tubes.

For an estimated film temperature of 90°F, a cleanliness factor judged to be 0.80, and a suitable water velocity of 6.5 fps, h_i[by Eq. (7-5)] = 1242 Btu/(hr)(sq ft)(°F) and

$$\frac{D_o}{D_i h_i} = \frac{0.75}{0.652 \times 1242} = 0.000926 \ (\text{hr})(\text{sq ft})(°F)/\text{Btu}$$

6. The over-all thermal resistance for operating conditions will now equal the sum of the values found for the separate resistances in steps 3, 4, and 5. Thus,

$$\frac{1}{U_o} = 0.000067 + 0.000964 + 0.000926 = 0.001957 \ (\text{hr})(\text{sq ft})(°F)/\text{Btu}$$

and
$$U_o = 512 \ \text{Btu}/(\text{hr})(\text{sq ft})(°F)$$

No account has yet been taken of the rate of condensation. When this has been determined, the values of h_o and U_o will probably need to be revised.

7. In order that the mean temperature difference may be predicted, the outlet-water temperature must be estimated. For this purpose it may be observed that for test conditions the rise in temperature of the cooling water, viz., $87.0 - 73.3$, or $13.7°$, was 63.2 per cent of the difference between the temperatures of the steam and the inlet water, which was $95.0 - 73.3$, or $21.7°$. Under specified operating conditions the difference between the steam and the inlet-water temperatures will equal $101.14 - 80$, or $21.14°$. For the same ratio of 63.2 per cent, the outlet-water temperature would be $80 + 13.3$, or $93.3°$; but because of the increase in water velocity under specified operating conditions a smaller rise in water temperature should be expected. For a first approximation an outlet-water temperature of $92°$ may be assumed. The mean temperature difference will then be

$$\frac{(101.14 - 80) - (101.14 - 92)}{\ln\left[(101.14 - 80)/(101.14 - 92)\right]} = 14.33°\text{F}$$

The heat transferred per square foot of outer surface will be

$$14.33 \times 512 = 7335 \text{ Btu/(hr)(sq ft)}$$

The corresponding loading will be

$$\frac{7,335}{950} = 7.73 \text{ lb/(hr)(sq ft)}$$

Reference to Fig. 12-2 will show that at a loading of 7.73 lb/(hr)(sq ft) the loading factor is 0.99, whereas at the loading of 7.22 lb/(hr)(sq ft), which prevailed during the test, the loading factor is 0.975. The value of h_o as found in step 4 should therefore be increased in the ratio of 0.99/0.975. The revised value of h_o will be $1037 \times (0.99/0.975) = 1053$ Btu/(hr) (sq ft)(°F). The revised value of U_o will be 515 Btu/(hr)(sq ft)(°F), and the revised rate of the heat transfer will be 7380 Btu/(hr)(sq ft), which corresponds to a loading of $7,380/950 = 7.77$ lb/(hr)(sq ft). The increase in loading from 7.73 to 7.77 lb/(hr)(sq ft) is apparently too small to require any further revision in h_o and U_o.

At a loading of 7.77 lb the extent of outer surface required to condense 33,000 lb of steam per hour will be $33,000/7.77 = 4,250$ sq ft.

Table 12-1, taken from manufacturer's data, shows the maximum desirable tube length for standard two-pass condensers of various sizes.*

According to this table the maximum desirable tube length for $\frac{3}{4}$-in. tubes is 15 ft, which will be assumed for further calculation of the problem. The outer surface of a $\frac{3}{4}$-in. by 15-ft tube is 2.94 sq ft. The required number of tubes, therefore, is $4,250/2.94 = 1,446$, and the number in each of the two passes is 723.

The inside cross-sectional area of 723 tubes (18 BWG) is

$$723 \times 0.00231 = 1.670 \text{ sq ft}$$

---

* See also Ref. 6.

The quantity of water that will pass through an area of 1.670 sq ft, at a velocity of 6.5 fps, will be $6.5 \times 1.670 = 10.85$ cu ft/sec, or 39,060 cu ft/hr, or (at an average temperature of about 85°F)

$$39,060 \times 62.14 = 2,427,200 \text{ lb/hr}$$

The rise in temperature of the water required for absorbing the specified quantity of heat will be $(33,000 \times 950)/2,427,000 = 12.90°F$. The outlet-

TABLE 12-1

Surface, sq ft	Maximum desirable tube length, ft		
	¾-in. tube	⅞-in. tube	1-in. tube
Up to 1,500	10	13	13
1,500– 2,000	13	15	15
2,000– 3,000	14	16	16
3,000– 6,000	15	17	17
6,000–10,000	18	18	20
10,000 and over	20	20	22

water temperature, accordingly, will be $80 + 12.90 = 92.90°F$ instead of the assumed value of 92°F. This discrepancy may be corrected by a second approximation.

By repeating step 7, assuming an outlet-water temperature between the two values, say 92.5°F, the mean temperature difference will be 13.99°F, and the rate of heat transfer will be $515 \times 13.99 = 7210$ Btu/(hr)(sq ft). The corresponding loading will be $7,210/950 = 7.59$ lb/(hr)(sq ft). The required surface will be 4,350 sq ft, and the number of tubes will be 1,480.

A repetition of the calculation of the temperature rise will show an outlet-water temperature of 92.65°F, which is sufficiently close to the assumed value of 92.50 to call for no further revision.

In commercial practice, modification in the computed number of tubes may be necessary in order that they may be suitably arranged. Provision for such a modification can usually be made by adjustment in the water velocity. The reader may observe also that in commercial practice the work involved in a problem of this sort may be greatly reduced by the use of graphical or tabular solutions applied to many of the necessary calculations.

CALCULATION OF HEAT TRANSFER THROUGH A
FIN-TUBE FAN COIL

12-6. General. A study of the heat transfer through fan coils of the fin-tube type will serve to illustrate the application of principles of heat transfer to a specialized field.

Coils of this type are commonly employed in fan systems

designed for supplying heated, cooled, or dehumidified air as desired in various applications of air conditioning. Fin-tube coils, as the name implies, are made of tubes surrounded by fins. The tubes are generally of copper or aluminum, and the fins either copper, brass, or aluminum. The fins are usually rectangular or circular in shape or are in the form of a helical ribbon, either smooth or crimped, wound onto the tube. Figure 12-3 shows several arrangements of tubes and fins.

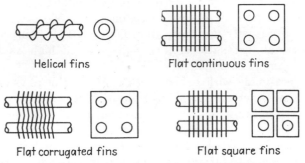

Helical fins   Flat continuous fins

Flat corrugated fins   Flat square fins

Fig. 12-3. Arrangement of tubes and fins.

Although there are many variations of design in special cases, the tubes are usually of small diameter, of the order of $\frac{3}{8}$ to $\frac{3}{4}$ in. o.d., spaced from $1\frac{1}{2}$ to 2 in. on centers. Fins are generally spaced from four to eight per inch of tubing. Good contact for the conduction of heat from the tubes to the fins, or in the opposite direction, is assured in many instances by dipping the assembled unit in hot tin or zinc.

Calculations of heat transfer through fin-tube coils differ from the calculations heretofore encountered primarily because of the marked difference between the extent and shape of the exterior and the interior surfaces. Furthermore, where a coil is used for cooling to a temperature below the dew point of the incoming air, the rate of heat transfer is affected by the film of condensate that covers a part or all of the exterior surface.

The exterior surface of the *tubes* is known as *primary* surface, and the *fin* surface is called *secondary* surface. The extent of secondary surface is in most cases such as to make the ratio of outside to inside surface within the range of from 10:1 to 30:1.

The tubes in any unit are of such number, length, and spacing as to provide the necessary area of opening for the desired quantity of air to flow through at a suitable velocity. The average velocity of the air flowing over the surface of the coil is commonly expressed as the average velocity over the entire face area, i.e., the volume of air per unit of time divided by the entire area of the face of the coil. In studies of heat transfer, however, it is often desirable to express the velocity in terms of the net or free air opening between the fins and tubes, i.e., the entire face area minus the projected area of the fins and tubes.

Face velocities for coils used in public buildings or in other places where the noise that accompanies high velocities is objectionable are usually not more than 600 or 700 ft/min; whereas for industrial installations, face velocities as high as 800 to 1,200 ft/min are common.

Units are made with one or with a number of rows of tubes in depth in order to provide the desired degree of heating or cooling, the limiting factor being the increase in frictional resistance to air flow that accompanies the increase in depth.

The basis for the use of fins on the outside of heat-transfer tubing is apparent. In the passage of heat through a tube filled with a liquid or vapor and surrounded by air, the greatest thermal resistance encountered is on the outer, or air, side. It is logical to expect, then, that this resistance can be lessened by increasing the exterior surface, as by the addition of fins, provided other factors affecting the surface resistance are not greatly altered.

12-7. Heat Transfer of Finned Tubing. The major factors affecting the heat transfer of finned tubing are as follows:

1. The mean velocity of the air stream.

2. The degree of scrubbing action of the air stream as influenced by the design and arrangement of the tubes and fins and by the presence of eddy currents or local turbulence.

3. The condition of the exterior surface, wet or dry.

4. The thermal conductivity, depth, thickness, and spacing of fins and the method of attaching them to the tube surface.

5. The nature and velocity of the fluid inside the tubes.

6. Other factors of lesser influence are the diameter of the tubes; the direction of heat flow, i.e., for heating or for cooling; and the absolute temperature of the air.

Fin-tube units are so compact that the enveloping or radiating surface per unit of heat capacity is small. By comparison with the flow of heat by convection, radiation can be of little consequence in any study of the performance of finned tubes, and it is usually considered negligible. Likewise, the resistance to heat flow through the thin metal tube itself is so small in comparison with the surface resistances that it may well be left out of consideration and the problem may be considered to deal with heat flow by convection only. The over-all coefficient U_o may then be expressed in terms of the inside- and outside-surface conductances.

$$U_o = \frac{1}{(R/h_i) + (1/h_o)} \tag{12-5}$$

where h_i = inside- (or refrigerant-side) surface coefficient.

h_o = effective outside- (or air-side) surface coefficient.*

R = ratio of the exterior to the interior surface.

U_o = over-all coefficient of heat transfer per square foot of exterior or air-side surface per degree Fahrenheit difference in temperatures of the fluid inside the tube and of the air.

The over-all coefficient U_o can readily be determined experimentally for any certain set of conditions by measurement of all remaining values expressed in the usual simple equation:

$$q = A U_o \Delta t_m \tag{12-6}$$

where q = rate of heat transfer, Btu/hr.

A = area of the exterior, or air-side, surface, sq ft.

U_o = over-all coefficient as defined above.

Δt_m = mean temperature difference (usually logarithmic) between the heating or cooling fluid inside the tube and the air.

The total heat transfer may be determined as the product of the pounds of heating or cooling medium condensing or evaporating within the tubes, or flowing through the tubes, per unit of time and the heat removed from or added to each pound of fluid. Otherwise, the measurements may be made on the air side, and the total heat determined as the weight of air per unit of time multiplied by its specific heat and by its change in temperature. In the case where air is cooled and condensation of

---

* The effective outside-surface coefficient as here applied is in reality the product of the fin effectiveness and the outside-surface coefficient as it is usually defined.

water vapor occurs, the heat removed from the air is the product of the weight of air per unit of time and the difference in enthalpy per pound at the initial and final wet-bulb temperatures. Applications of the foregoing principles in a study of the performance of heat-transfer units are necessarily limited to coils of similar design and arrangement and with the same velocities of the air and the fluid inside the tubes as in the tests from which the heat-transfer coefficient has been determined.

12-8. Surface Coefficients. It should be remembered that the over-all coefficient U_o depends upon many variable factors already enumerated, and its value for various conditions of operation cannot be predicted from the results of a test at one set of conditions. In order, then, that its value may be computed for various conditions, knowledge of the surface coefficients h_i and h_o and the ratio of the external to the internal surface, R, of Eq. (12-5) is essential.

The surface coefficients h_i and h_o are not easily determined experimentally, since their determinations involve measurement of the temperature gradient, and that is difficult to accomplish. The difficulty arises (1) because of the differences in temperature between the external surface of the tubes and of the various portions of the fins and (2) because the inside-film conductance is comparatively great and the drop in temperature across this film is correspondingly small. Values of the over-all coefficient U_o, determined by test, however, can be broken down into the values of h_i and h_o and R, by tests at several refrigerant velocities and several air velocities or by computation of h_i and by measurement of the ratio R. Substitution of these values of U_o, h_i, and R in Eq. (12-5) establishes the value of the effective outside-surface coefficient h_o. The logic of the latter procedure lies in the fact that fewer variables affect the inside-film conductance than the outside-film conductance, and their influence is in most cases well established. Further, since the thermal resistance of the inside film is commonly much less than that of the outside film, the effect of any inaccuracies in the computation of h_i upon the outside coefficient h_o is minimized.

The inside-surface coefficient for boiling liquids may be determined by application of Eq. (9-3) and Tables 9-2 and 9-3, with values increased by some 25 per cent to allow for the effect of the tubular shape of the surface. Otherwise, in the case of the common refrigerants dichlorodifluoromethane (Freon-12) and methyl chloride, the approximate value of 300 Btu/(hr)(sq ft)(°F) may be assumed.

The inside-surface coefficient for condensing steam in fan coils is lower than the corresponding coefficient for steam condensers, because of the slower drainage of condensate from the inner surfaces of the tubes. The common value of 2000 Btu/(hr)(sq ft)(°F), ordinarily applied to steam condensing on a single tube, is accordingly reduced to about 1200 when applied to the design of fan coils.

Values of the inside-surface coefficient may be computed satisfactorily by the methods outlined in Chaps. 7, 9, and 10. More specifically, where the heating or cooling medium inside the tubes does not change its state from liquid to vapor or from vapor to liquid, Eq. (7-5) may be applied. For *water* at temperatures not exceeding 180°F, Eq. (7-14) is applicable. Tests have been reported by William Goodman[9] showing inside-surface coefficients 30 per cent above values computed by Eq. (7-5) or (7-14); but since these tests were made on tubes enclosing aluminum spirals, inserted for the purpose of increasing turbulence, the results verify rather than deny the suitability of Eqs. (7-5) and (7-14) for usual application.

The following example will illustrate the determination of the surface coefficients h_i and h_o in a specific case where the over-all coefficient U_o has been established by test.

Example 12-3. Tests of a certain fin-tube coil used to cool dry air show an over-all coefficient U_o of 5.0 Btu/(hr)(sq ft)(°F) when water at an average temperature of 50°F is the cooling medium. The water flows through tubes of 0.375 in. i.d. at a velocity of 2 fps. The ratio of external to internal surface, R, is 20:1. Determine the surface coefficients h_i and h_o.

By Eq. (12-5)

$$U_o = \frac{1}{(R/h_i) + (1/h_o)} = 5.0 \text{ Btu/(hr)(sq ft)(°F)}$$

By Eq. (7-14)

$$h_i = 0.00134(t + 100)\frac{V^{0.8}}{D^{0.2}}$$

For test conditions

$$t = 50, \qquad D^{0.2} \text{ (by Table 7-3)} = 0.50$$
and
$$V^{0.8} = (3600 \times 2)^{0.8} = 1218$$

by Eq. (7-14)

$$h_i = 0.00134 \times 150 \times \frac{1218}{0.500} = 490 \text{ Btu/(hr)(sq ft)(°F)}$$

By substitution of the foregoing values into the equation

$$U_o = \frac{1}{(R/h_i) + (1/h_o)} \quad 5.0 = \frac{1}{(20/490) + (1/h_o)}$$

and $\quad h_o = 6.28 \text{ Btu}/(\text{hr})(\text{sq ft})(°F)$

For a graphical method of determining values of h_i and h_o from test data, see Ref. 10.

The outside-surface coefficient and the fin effectiveness may be determined from the plots established by Kays and London, as described in Art. 7-23. Otherwise, where coils are to be used for heating or cooling of air within the usual ranges of temperatures and where the only notable variables are the air velocities and the diameter of the tubes, the effective outside-surface coefficient may well be expressed in the form*

$$h_o = \frac{CG^n}{D^{1-n}} \tag{12-7}$$

where h_o = effective outside-surface coefficient, Btu/(hr)(sq ft of total air-side surface)(°F difference between the mean air temperature and the air-side surface temperature).

C and n = constants that must be determined by experiment.

D = outside diameter of the tubes, ft.

G = mass velocity, $\text{lb}_m/(\text{hr})(\text{sq ft of face area})$ or (sq ft of net area). (C will vary according to which basis is adopted.)

Tuve and McKeeman[11] have summarized values of the exponent n, as determined by various experimenters, and have found these values to range from 0.4 to 0.8; but from their own experimental work they have observed that the values of n are approximately 0.6 in the case of common large fin tubes in staggered arrangement and 0.5 in the case of small tubes with large fins. Corresponding values of C are approximately 0.04 in the former case and 0.105 in the latter. The mass velocity G is here expressed as the pounds of air flowing per hour per square foot of net or free area between the fins and tubes.

For accurate results, the values of C and n should be determined by tests of a coil at several air velocities. When these values have been established for a coil of any certain design, the performance may be satisfactorily predicted for coils of similar form, of any desired dimensions, i.e., for any length of coil with any number of rows of tubes in width and depth.

The following example will illustrate the application of the foregoing principles to the design of a coil.

---

* It may be noted that this equation is in the general form of Grimison's equation, Eq. (7-24), except that C has replaced $B(k/\mu^m)$. This comparison shows the limitation that must be placed upon Eq. (12-7), namely, that it is applicable to a specific fluid in a limited range of temperature where little variation occurs in the value of k/μ^m.

Example 12-4.   A fin-tube fan coil is to be designed to heat 4,000 cu ft of air per minute from 80 to 160°F.   The heating medium is to be steam at a pressure of 5 psig (227°F).   Tubing is to be of ½-in. i.d. and ⅝-in. o.d., and the ratio of the outer to the inner surface will be 16:1.   Tests of a coil of similar type show the outside-surface coefficient to be represented by the equation

$$h_o = 0.041 \frac{G^{0.5}}{D^{0.5}} \qquad \text{[in the form of Eq. (12-7)]}$$

where G = mass velocity in $\text{lb}_m/(\text{hr})(\text{sq ft of } \textit{net air opening})$ and D is the outside diameter of the tube in feet.

The net air opening in this type of coil is 60 per cent of the face area.

Solution.   The initial air density (Table A-2) equals 0.0734.   The amount of heat to be added to the air equals

$$4000 \times 60 \times 0.0734 \times 0.24(160 - 80) = 338{,}227 \text{ Btu/hr}$$

At a chosen air velocity of 600 fpm through the face area, the face area will equal $4{,}000/600 = 6.66$ sq ft.

The net air opening will be $0.60 \times 6.66 = 4.0$ sq ft, and the velocity through the net air opening will be $600/0.60 = 1{,}000$ fpm.

The effective outside-surface coefficient, according to the formula derived from test data, will be

$$h_o = 0.041 \frac{(0.0734 \times 1000 \times 60)^{0.5}}{(0.625/12)^{0.5}}$$

$$= 0.041 \times \frac{4404^{0.5}}{0.052^{0.5}}$$

$$= 0.041 \frac{66.4}{0.228}$$

$$= 11.9 \text{ Btu/(hr)(sq ft)(°F)}$$

The inside-surface coefficient h_i for condensing steam may be taken to be 1200 Btu/(hr)(sq ft)(°F).

$$U_o = \frac{1}{(R/h_i) + (1/h_o)} = \frac{1}{(16/1200) + (1/11.9)} = \frac{1}{0.0133 + 0.0840}$$

$$= \frac{1}{0.0973} = 10.28 \text{ Btu/(hr)(sq ft)(°F)}$$

The temperature difference between steam and air will be

$$\frac{(227 - 80) - (227 - 160)}{\ln\left[(227 - 80)/(227 - 160)\right]} = 101.9°F$$

The amount of heat transferred per hour per square foot of external surface equals $U_o \, \Delta t_m = 10.28 \times 101.9 = 1047.5$ Btu.   The amount of external surface needed will be $338{,}227/1047.5 = 323$ sq ft.   The dimensions of the coil will be such as to provide the desired 6.66 sq ft of face area with the number of rows of tubing in depth sufficient to provide, as nearly as possible, the 323 sq ft of external surface.

12-9. Coils Operating with Dehumidification. So far, the discussion has dealt mainly with heat transfer through coils with the air side in a dry condition. When coils are used to cool air to temperatures below the dew point, the external surfaces of the coil are covered, either partially or completely, with a film of condensate, and consequently a number of additional factors are involved in a study of the heat transfer.

The designer of a coil to be used for cooling and dehumidification is concerned not only with the total amount of heat to be removed from the air but also with the separate amounts of heat removed from the dry air (sensible heat) and from the water vapor (latent heat).

Knowledge of the proportionate amounts of sensible and latent heat is essential, since for certain applications the main function of the coil is to remove moisture from the air, and in other cases the removal of moisture is incidental to the lowering of the air temperature. These varying requirements govern the design of the coil and the conditions under which it must be operated, particularly with regard to the temperature that must be maintained in the cooling medium.

Many methods have been devised for predicting the performance of a dehumidifying coil. These methods in some cases are largely empirical or are based upon heat-transfer theory to a limited extent. In some cases calculations are based upon the over-all coefficient U_o for dry coils of fixed depth and for fixed velocities of the air and the cooling medium, with correction factors applied in the case of wet coils in accordance with the temperature differences or according to the ratio of the latent- to the sensible-heat load.

Several methods, which appear more rational, have applied heat-transfer principles to the extent of making use of surface coefficients or of basing calculations upon the average temperature of the external surface of the coil. A method devised by William Goodman[9] by a rigid mathematical development makes use of surface coefficients as functions of the velocities of the air and the cooling medium, although it also requires the use of a number of empirical factors. In the methods that involve the average temperature of the external surface, that temperature is rarely found directly by test measurements; it is more commonly a calculated value determined from other test data,[10] or it is a fictitious value found graphically.[12]   The graph is drawn on a (nonlogarithmic) psychrometric chart, and the surface temperature determined as the point of intersection of the saturation curve with a straight line starting from the point representing the condition of the air that enters the coil and extending at an angle that corresponds to the ratio of sensible to total heat load.

No attempt is made here to describe the various methods in detail.   For such descriptions the reader is referred to the references already cited and to a paper by A. P. Colburn and O. A. Hougen on the Design of Cooler Condensers for Mixtures of Vapors with Noncondensing Gases.[13]   By Goodman's and by Colburn's and Hougen's methods it is possible to predict the progressive changes in the condition of the air and vapor mixture as it passes each row of tubes in a coil, although the calculations required by these methods are somewhat involved and extensive.

The main result that is ordinarily desired, however, is the final condition of the air as it leaves the coil.   Even though the temperature of the external surface of the coil is below the dew-point temperature of the entering air, and water vapor is removed by condensation, complete saturation of the air leaving the coil is seldom obtained, since not all the air flowing over the coil comes into intimate contact with the cold surface.   The air leaving the coil is in reality a mixture of air which has been cooled and dehumidified by contact with the cold surface and air which has bypassed the surface without reaching the dew-point temperature.   In ordinary practice the surface of a fin-tube coil four rows in depth is contacted by approximately 80 per cent of the air, whereas the surface of a six-row coil is contacted by approximately 95 per cent of the air.[14]

The final condition of the air can be predicted in a comparatively simple manner by a method developed by G. L. Tuve.[15] It is this method which will be applied here in the further study of the performance of coils used for dehumidifying air.

In Tuve's method the outside-surface temperature is calculated and applied as if the outside surface were entirely prime surface (i.e., without fins).   Surface coefficients are obtained from over-all coefficients in the manner already explained in the case of dry coils.   In fact, in this method the outside-surface coefficient is a coefficient of sensible heat only; and if found by tests, the tests must be conducted on a dry coil.   The relationship between the outside- (i.e., air-side) surface coefficient and the outside-surface temperature is expressed by the equation

$$h_o = \frac{0.243G}{AN} \ln \frac{t_1 - t_s}{t_2 - t_s} \qquad (12\text{-}8)$$

where h_o = outside-surface coefficient of sensible heat, Btu/(hr)(sq ft of air-side surface)(°F of logarithmic mean temperature difference between the dry-bulb temperature of the air and the outside-surface temperature).

G = mass velocity of the air, lb_m/(hr)(sq ft of face area).

A = outside-surface area, sq ft/(sq ft of face area) (row of coil depth).

N = number of rows of coil depth.

t_1 = entering dry-bulb temperature, °F.

t_2 = exit dry-bulb temperature, °F.

t_s = outside-surface temperature, °F.

0.243 = specific heat of humid air, Btu/(lb_m)(°F).

NOTE: Equation (12-8) is derived by expressing the sensible heat transfer per square foot of face area, $c_p G(t_1 - t_2)$ equal to $h_o \, \Delta t_m \, AN$, where the mean temperature difference

$$\Delta t_m = \frac{t_1 - t_2}{\ln \left[(t_1 - t_s)/(t_2 - t_s)\right]} \qquad \text{°F}$$

A feature of Tuve's method, which is found in no other, is the manner of finding the condition of the air leaving the coil when dehumidification is taking place. Tuve has shown by mathematical analysis and by test results that the difference between the dry-bulb temperature and the dew point of the air leaving a wet coil is substantially the same as the difference between the dry-bulb temperature and the dew point of the air leaving the same coil when it is dry but with its surface at the same temperature as the dew point of the entering air.

The procedure for finding the condition of the exit air involves first the calculation of the dry-bulb temperature of the exit air when the surface is at the dew-point temperature of the entering air. This temperature is readily found by substitution of appropriate values in Eq. (12-8), solving for t_2. The intersection of this dry-bulb temperature (vertical line) on a psychrometric chart of the nonlogarithmic type with the line of constant moisture content of the entering air (horizontal line) establishes the condition of the exit air.

Figure 12-4 is a portion of a nonlogarithmic psychrometric chart in which the point a represents the condition of the air entering the coil. Point b represents the condition to which the air would be cooled if the outside surface were maintained at the dew-point temperature of the entering air. It is located on the psychrometric chart at the intersection of the horizontal (constant moisture content) line through point a with the vertical line drawn from the temperature t_2 [t_2 being found by Eq. (12-8)].

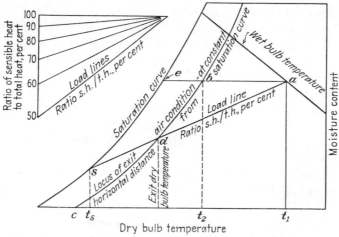

FIG. 12-4. Psychrometric chart showing method of predicting coil performance.

The condition to which the air will be cooled when dehumidification is taking place will lie somewhere on the line bc, a line at a constant distance from the saturation curve. The exact point will depend upon the heat-load ratio which the coil is to maintain, i.e., the ratio of the sensible heat to the total heat removed from the air. This ratio can readily be represented on the psychrometric chart of the nonlogarithmic type by a straight line of definite slope; for in this form of chart equal increments on the abscissa (dry-bulb temperature) represent equal amounts of sensible heat, and equal increments on the ordinate (moisture content) represent substantially equal amounts of latent heat. If this heat-ratio line, when extended, intersects the saturation curve, the point of intersection indicates the temperature of the outside surface of the coil. It is obvious, therefore, that the ratio of sensible heat to total heat maintained by a cooling coil is dependent upon the outside-surface temperature, which, in turn, is dependent upon the temperature of the refrigerant in the coil. Figure 12-4 shows, in the upper left portion, a set of master slope lines corresponding to various heat-load ratios. If a line ad of such a slope as to represent the desired heat-load ratio is drawn on the psychrometric chart, starting at the point a which represents the inlet-air conditions, its intersection d with the curved line bc at a distance be from the saturation curve will establish the exit dry-bulb temperature at the point d.

The surface temperature required for maintaining the desired exit dry-bulb temperature may now be found by again applying Eq. (12-8), this time with t_2 equal to the dry-bulb temperature at point d.

Since both the inlet and outlet dry-bulb temperatures are now known, the sensible heat removed from the air per hour may now be determined as the product of the specific heat of the air, its mass in pounds per hour, and the change in dry-bulb temperature.

The refrigerant temperature may be found by solving the following equation for the value of t_r.

$$q_t = \frac{h_i}{R} (t_s - t_r) AN \qquad (12\text{-}9)$$

or

$$t_r = t_s - \frac{Rq_t}{h_i AN} \qquad (12\text{-}10)$$

where q_t = total heat transfer, Btu/hr.

h_i = refrigerant- (inside-) surface coefficient, Btu/(hr) (sq ft)(°F).

R = ratio of air-side to refrigerant-side surface.

t_s = surface temperature, °F.

t_r = refrigerant temperature, °F.

A = air-side surface, sq ft/(sq ft of face area)(row of coil depth).

N = number of rows of coil depth.

The derivation of this expression is obvious when it is observed that AN/R is the inside, or refrigerant-side, surface in square feet per square foot of face area. With these values established, a variety of problems relating to the performance of a dehumidifying coil may be solved.[10,15]

The following example will illustrate the procedure when the condition of the entering air together with air and refrigerant velocities and the heat-load ratio are known and it is desired to find the condition of the exit air; the sensible, the latent, and the total heat transfer; and the refrigerant temperature.

Example 12-5.   A fan coil made of fin tubes five rows deep has a face area of 3.75 sq ft, an outside surface of 282 sq ft, and an inside tube surface of 22.9 sq ft. It is to be used to cool and dehumidify 1,500 cu ft of air per minute entering the coil at 80° dry-bulb and 67° wet-bulb temperatures. The sensible-heat load is to be 70 per cent of the total heat transfer. The cooling medium is cold water flowing through ⅝-in. i.d. tubes at a velocity of 2 fps. Tests of the coil with air at the given conditions of dry-bulb temperature and velocity and with the outside surface in a dry condition have shown the outside-surface coefficient h_o to be 3.90 Btu/(hr)(sq ft of outside surface) (°F).

Determine the wet- and dry-bulb temperatures of the air leaving the coil; the sensible, the latent, and the total heat transfer; and the refrigerant temperature.

Solution.   The dew point of the entering air, as read from a Carrier psychrometric chart, is 60.3°F. The dry-bulb temperature of the exit air that would prevail if the surface temperature t_s were 60.3°F is established by Eq. (12-8) by solving for t_2 as follows:

$$h_o = \frac{0.243G}{AN} \ln \frac{t_1 - t_s}{t_2 - t_s} = 3.90$$

$$G = \frac{1,500 \times 60 \times 0.0731}{3.75} = 1,750 \text{ lb}_m/(\text{hr})(\text{sq ft of face area})$$

$$A = \frac{282}{3.75 \times 5} = 15.04 \text{ sq ft}$$

$$N = 5$$

$$t_1 = 80$$

$$t_s = 60.3$$

$$3.90 = \frac{0.243 \times 1,750}{15.04 \times 5} \ln \frac{80 - 60.3}{t_2 - 60.3}$$

$$t_2 = 70.2°\text{F}$$

The temperature difference, $t_2 - 60.3 = 70.2 - 60.3 = 9.9°F$, establishes the locus of the exit air temperature t_2 on the psychrometric chart at a horizontal distance of 9.9°F from the saturation curve. The intersection of this locus and a 70 per cent load-ratio line drawn from the point representing the inlet air conditions shows the exit air to be at a dry-bulb temperature of 66.6°F and a corresponding wet-bulb temperature of 60.6°F.

The sensible-heat transfer in Btu/(hr)(sq ft of face area) in cooling from 80 to 66.6° will be

$$G \times 0.243 \times (80 - 66.6) = 1,750 \times 0.243 \times 13.4$$
$$= 5700 \text{ Btu/(hr)(sq ft of face area)}$$

The total heat transfer per square foot of face area, for the 70 per cent load ratio, will be

$$\frac{5700}{0.70} = 8130 \text{ Btu/(hr)(sq ft of face area)}$$

The latent-heat transfer $= 8130 - 5700 = 2430$ Btu/(hr)(sq ft of face area).

The surface temperature may now be found by again applying Eq. (12-8).

$$h_o = \frac{0.243G}{AN} \ln \frac{t_1 - t_s}{t_2 - t_s}$$
$$3.90 = \frac{0.243 \times 1,750}{15.04 \times 5} \ln \frac{80 - t_s}{66.6 - t_s}$$
$$t_s = 53.4°\text{F}$$

The refrigerant temperature as found by Eq. (12-10) will be

$$t_r = t_s - \frac{Rq_t}{h_i AN}$$

All values on the right-hand side of this equation are known, with the exception of the refrigerant-side surface coefficient h_i. The value of h_i for water at a velocity of 2 fps in a ⅝-in. i.d. tube, at a temperature of approximately 50°F, may be found by Eq. (7-14) as follows:

$$h_i = 0.00134(100 + 50) \frac{(3600 \times 2)^{0.8}}{[5/(8 \times 12)]^{0.2}}$$
$$= 442 \text{ Btu/(hr)(sq ft)(°F)}$$

NOTE: Inaccuracy in the assumption of the water temperature for determining h_i will have little effect upon the final result.

By Eq. (12-10)

$$t_r = 53.4 - \frac{282 \times 8130}{22.9 \times 442 \times 15.04 \times 5} = 50.4°\text{F}$$

Summarizing, the desired temperatures will be

Exit dry-bulb temperature $= 66.6°$F
Exit wet-bulb temperature $= 60.6°$F
Refrigerant temperature $= 50.4°$F

and the heat transfer for the entire coil will be

Sensible-heat transfer $= 3.75 \times 5700 = 21,375$ Btu/hr
Latent-heat transfer $= 3.75 \times 2430 = 9,113$ Btu/hr
Total heat transfer $= 3.75 \times 8130 = 30,488$ Btu/hr

In problems where the desired condition of the exit air has been established, features of the design of a coil and its rating may best be determined by expressing Eq. (12-8) in a more advantageous form:

$$h_o = \frac{0.243G}{AN} \ln \frac{t_1 - t_{dp1}}{t_2 - t_{dp2}} \qquad (12\text{-}11)$$

where t_{dp1} and t_{dp2} = dew-point temperatures of the air entering and leaving the coil.

The equality of the terms $(t_1 - t_s)/(t_2 - t_s)$ in Eq. (12-8) and $(t_1 - t_{dp1})/(t_2 - t_{dp2})$ in Eq. (12-11) may be observed by examination of Fig. 12-4 in which the triangular figures aes and abd are similar. The ratio of lengths $as/ds = ae/be$, or, in terms of temperature differences, $(t_1 - t_s)/(t_2 - t_s) = (t_1 - t_{dp1})/(t_b - t_{dp1})$; but since all points on the line bc are at a constant horizontal distance from the saturation curve es, $t_b - t_{dp1}$ may be replaced by its equivalent $t_2 - t_{dp2}$, which results in the expression $(t_1 - t_s)/(t_2 - t_s) = (t_1 - t_{dp1})/(t_2 - t_{dp2})$.

NOTE: This derivation is not strictly accurate because of the deviation of the saturation curve from a straight line connecting points e and s. The inaccuracy, however, is not so great as to invalidate the conclusion that as the outside-surface temperature of a coil is changed, the difference between the dry-bulb temperature of the exit air and its dew-point temperature remains substantially constant.

Example 12-6. A fin-tube coil is to be designed to cool air at an initial condition of 95°F dry-bulb and 76°F wet-bulb to a final condition of 60°F dry-bulb and 57°F wet-bulb temperature. The coil is to be designed for an inlet air velocity over the face area of 400 fpm, at which tests of a similar coil have shown an outside-surface coefficient h_o of 9.5. The outside-surface area A of this type of coil is 16 sq ft/(sq ft of face area) (row of coil depth).
Determine the number of rows of coil depth.
Solution. At inlet-air conditions,

$$G = 400 \times 60 \times 0.0709 = 1{,}702 \text{ lb}_m/(\text{hr})(\text{sq ft of face area})$$

$t_1 = 95°F$, and for a wet-bulb temperature of 76°F the dew-point temperature $t_{dp1} = 68.4°F$.
At exit-air conditions, the dew-point temperature $t_{dp2} = 55.1°F$.
The number of rows of coil depth may be found by substitution in Eq. (12-11) as follows:

$$h_o = \frac{0.243G}{AN} \ln \frac{t_1 - t_{dp1}}{t_2 - t_{dp2}} = 9.5$$

$$\frac{0.243 \times 1,702}{16 \times N} \ln \frac{95 - 68.4}{60 - 55.1} = 9.5$$

$$N = 4.60$$

Since the value of N lies between 4 and 5 the choice of either four or five rows of coil depth should be based upon the results found by substituting in Eq. (12-8) the values for N of 4 and 5, solving for $t_2 - t_s$. These substitutions will show that when $N = 4$, $t_2 - t_s = 6.1°F$, and when $N = 5$, $t_2 - t_s = 4.2°F$.

The extension of the load-ratio line on the psychrometric chart from inlet conditions to exit conditions and to the saturation curve shows an outside-surface temperature of 49.8°F.

For the same outside-surface temperature, the exit-air conditions for four rows, as read from the psychrometric chart, will be 62.2°F dry-bulb temperature and 58.4°F wet-bulb; and for five rows they will be 58.8°F dry-bulb and 56.2°F wet-bulb. The closer approach to the desired exit-air conditions, therefore, is found with five rows of coil depth.

The desired exit-air conditions may be attained more exactly with either four or five rows of coil depth by resorting to one or more of the following means: by modification of the air velocity which in turn alters the value of h_o; by modification of the refrigerant temperature and consequently the outside-surface temperature; by resorting to reheat; by mixing return air with the incoming air; and by mixing the exit air with air that is bypassed around the cooling coil.

CALCULATION OF HEAT TRANSFER IN AN ELECTRICAL TRANSFORMER

12-10. General. The designs and capacities of many forms of electrical equipment, such as motors, generators, transformers, heaters, and electronic devices, are governed to a great extent, and in some cases primarily, by the means for the removal of the heat which is generated within the equipment. To illustrate the application of the principles of heat transfer to one such form, the following example relating to the design of electrical transformers has been selected. This example will serve also to show a practice, common in electrical-engineering literature, of using a system of units which, although far from being consistent, is nevertheless in terms most familiar to the electrical engineer.

Transformers are installed most commonly where changes in voltage and current in a.c. circuits are desired. Less frequently their purpose is to change the number of phases or the phase relationships of the circuits or the separation of a grounded from an ungrounded circuit. Because of the heating effect due to the iron and copper losses which occur within the transformer, the rating at which it may be operated is dependent upon the effectiveness of the means provided for the removal of heat.

In the various designs of transformers, the medium for removing heat from the windings and the core may be air, moved by natural or by forced circulation, or it may be an oil or a synthetic liquid in which the windings and core are immersed. Heat is also removed from the exterior surface of the transformer by radiation and by natural or by forced convection of air. In oil-immersed transformers the oil may serve primarily as a means of transferring heat from the core and windings to the metal tanks in which the transformer elements are housed. If the heat transfer from the tank is insufficient, the rate of heat removal from the oil may be increased by natural or by forced circulation of the oil through tubes connected to the outer surface of the transformer or by pumping the oil through a separate air- or water-cooled heat exchanger. Otherwise the oil may be cooled by a water-cooled coil submerged in the oil within the transformer.

The choice of the various arrangements for heat removal is influenced by many considerations, such as the size of the transformer, the initial cost, the anticipated location either indoors or outdoors, the cost of water and the danger of its freezing, and the cost of power and maintenance if forced circulation of air, oil, or water is employed. The following comments apply to some of these considerations.

In the flow of heat from the transformer elements to its ultimate point of removal, the greatest resistances occur at the surfaces in contact with air. The simplest means of reducing the inside-surface resistance is by immersing the transformer elements in oil or other suitable liquid. If the exterior of the tank is a plain smooth surface, an insufficient amount of heat will be transferred to the surroundings by radiation and by free convection of air except in the case of transformers of small capacity. Above 25-kva size it is uneconomical to provide smooth tanks with large enough surface to dissipate the required amount of heat. The exterior surface of a tank may, however, be increased several fold by means of corrugations, without increase in transformer volume. The heat transfer by free convection from the corrugated sides of such a tank is increased somewhat less than in direct proportion to the increase in surface, because of air friction. The effect of friction must be determined experimentally since it depends upon the depth and spacing of the corrugations. The heat transfer by radiation is substantially the same as from a tank of equal volume with plain surfaces. When greater heat transfer than can be economically provided by radiation and free convection from corrugated surfaces is required, resort must be made to forced-air cooling, to external metal tubes, or to separate heat exchangers for cooling the oil or to water-circulating coils submerged in the oil.

12-11. Dry-type Self-cooled Transformers. Dry-type transformers cooled by free convection of air are generally limited to small capacities at low voltages for installation in dry places. Some transformers of this type with corrugated surfaces, however, have been marketed in sizes up to 500 kva, single-phase, and 300 kva, three-phase, and voltages up to 13,800.

12-12. Dry-type Forced-air-cooled Transformers. Dry-type forced-air-cooled transformers, arranged with fans mounted under the cores and windings, have been built in sizes up to 18,500 kva and 15,000 volts. Several transformers are sometimes installed over a common air chamber or air duct with dampered connections to each transformer. A supply of clean (usually filtered) air is required in an amount of approximately 150 cu ft/min/kw of total loss—a rate based upon an air temperature rise of about 21°F. In order to provide high rates of heat transfer, clearances for air passage are kept small, with the result that air-resistance pressures are of the order of $\frac{3}{4}$ in. of water for the smaller units and up to nearly 7 in. of water for the larger units. The increasing practice of installing transformers out of doors and the general increases in size and voltage have led to the almost complete abandonment of this type for new installations. For power-system installations oil-immersed transformers are almost universal.

12-13. Oil-immersed Self-cooled Transformers. Oil-immersed transformers, cooled by radiation and by free convection of air, are by far the most common type. This type of construction is usually justifiable even if at increased initial cost, because of the absence of auxiliary cooling equipment and its maintenance. In the smaller sizes, the cooling surface of the tank may be adequate, especially if it is in corrugated shape; but in most cases it is augmented by round or flattened vertical tubes welded to the outside of the tank with openings into the tank near the top and bottom so as to provide oil circulation by natural convection. In sizes above 7,000 kva, single-phase, or 10,000 kva, three-phase, the external tubes are connected into top and bottom headers, which are in turn bolted to flanged openings in the tank. The radiators which are thus formed may be placed in various ways around the tanks, according to the available space.

12-14. Oil-immersed Forced-oil-cooled Transformers. A practice which has found limited acceptance in the United States but considerable favor in Europe is that of pumping the oil from the transformer tank through an external oil-to-water or an external oil-to-air heat exchanger. Within the transformer tanks, the oil may be directed over the windings at increased velocity. With these means of providing efficient heat transfer to and from the oil, a reduction in the size and weight of the transformer may be effected.

12-15. Oil-immersed Forced-air-cooled Transformers. Fans are sometimes installed for cooling the exterior surface of oil-immersed transformers. These fans may be operated continuously or only during periods of heavy load. With such an arrangement two or more ratings are given to the transformer— one for self-cooling, another for complete fan operation, and sometimes a third which applies when some of the fans remain idle. The application of high-velocity fans to otherwise self-cooled transformers has increased their capacities by as much as 67 per cent.

12-16. Oil-immersed Water-cooled Transformers. Where an adequate supply of cooling water has been available at low cost, a practice sometimes followed, mostly in the United States, has been to circulate water through spirally wound copper coils submerged in the oil inside a smooth tank. Care has been taken to guard against leakage of water into the oil, by the use of jointless coils or by the maintenance of a water pressure below that of the oil. This method of cooling has effected a saving in initial investment, especially for large installations, but because of the care required to ensure continual circulation of the water, self-cooled transformers are generally preferred.

12-17. Computations of Heat Transfer in Electrical Transformers. In the transfer of heat from the windings of an oil-immersed transformer, four main resistances are encountered. The first is the resistance of the insulation and oil film on the surface of the windings; the second is the resistance of the oil film on the inner surfaces of the tank and cooling tubes, if any; the third is the resistance of the metal tanks and/or tubes; and the fourth is the resistance to transfer by radiation and convection from the outer surfaces of the tank and tubes.

Measurements of the temperature gradient in transformers have made it possible to predict the resistance of the insulation and oil film on the surface of the windings, and from this prediction it is possible to approximate the relative values of all the resistances and thereby establish the complete temperature gradient.

The practice of recording the temperature rise of electrical apparatus in degrees centigrade, and the load in watts, has led to the custom of expressing heat transfer in electrical-engineering literature in terms of watts per square inch per centigrade degree of temperature rise. These units will be used in the following computations, together with corresponding expressions in consistent units of Btu/(hr)(sq ft) and degrees Fahrenheit.

12-18. Temperature Rise of Windings over Oil. The temperature rise of the windings over the oil is not uniform but varies from point to point along the path of the heat flow. The hottest spot in the windings is usually assumed to exceed the average temperature of the coils by not more than 10°C under rated load conditions. The temperature of the oil is likewise not uniform, but for normal loads the temperature of the top oil exceeds the average oil temperature by only approximately 2°C.

Tests of transformers have shown that within the usual range of transformer loadings, the temperature rise of the windings (either hottest-spot or average) over the adjacent oil can be calculated by an equation in the form

$$\theta = KW_c{}^n\mu^{n^1} \qquad (12\text{-}12)$$

where θ = temperature rise, °C.

K = a constant.

W_c = heat transfer, watts/sq in.

μ = absolute viscosity (see Table 12-2) at film temperature,* centipoises.

n = empirical constant, 0.70 to 0.85, depending on the
cooling conditions, but most commonly 0.75.

n^1 = an empirical constant of 0.25.

TABLE 12-2. VISCOSITY OF TRANSFORMER OIL
[*From Knowlton (ed.)*[16]]

Temperature		Viscosity		Temperature		Viscosity	
°F	°C	Centi-poises	lb$_m$/(ft)(hr)	°F	°C	Centi-poises	lb$_m$/(ft)(hr)
80	26.7	11.3	27.3	170	76.7	3.00	7.26
90	32.2	9.4	22.7	180	82.2	2.69	6.52
100	37.8	7.9	19.1	190	87.7	2.43	5.88
110	43.3	6.7	16.2	200	93.3	2.21	5.35
120	48.9	5.75	13.9	210	98.9	2.02	4.89
130	54.4	4.98	12.05	220	104.4	1.84	4.45
140	60.0	4.35	10.52	230	110.0	1.67	4.04
150	65.6	3.82	9.24	240	115.6	1.51	3.66
160	71.1	3.38	8.18	250	121.1	1.36	3.29

12-19. Heat Transfer from the Outer Surfaces. Heat transfer
from the outer surfaces of the transformer tank and from any
cooling tubes that may be attached to the tank is by the processes
of radiation and convection (usually free convection).

1. The heat transfer by radiation is commonly expressed in
electrical-engineering literature as

$$W_r = K\epsilon(T_1{}^4 - T_2{}^4) \qquad (12\text{-}13)$$

* Because of the temperature drop through the turn insulation, the film
temperature may be estimated to be slightly below the mean of the temperatures of the windings and the adjacent oil. Test data show film temperatures equal to the temperature of the adjacent oil plus about three-eighths
of the rise of the windings over the adjacent oil temperature. In problems
where the temperature of the oil is not known, it may be estimated and
eventually verified by a trial-and-error procedure.

where W_r = heat transfer, watts/sq in.

K = a constant, equal to 3.68×10^{-11}.

ϵ = emissivity factor.

T_1 and T_2 = hot body and ambient temperatures, °C abs.

If this equation is compared with the more general radiant-heat-transfer expression, $q = \sigma F_e F_A A_1 (T_1^4 - T_2^4)$, expressed in Btu/hr, it will be noted that in the development of Eq. (12-13) it has been assumed that the transformer is a small body completely enclosed, for which case the configuration factor is unity and the emissivity of the surfaces of the enclosure is disregarded. Conversion of the units shows agreement of the two expressions, except that Eq. (12-13) appears to have been based upon a value of 0.172×10^{-8} instead of 0.173×10^{-8} for the Stefan-Boltzmann constant.

2. Heat transfer by *free convection* is expressed in the electrical-engineering literature[16] by an empirical equation

$$W_c = 0.0014F\theta^n \tag{12-14}$$

where W_c is in watts per square inch; F is an experimentally determined air-friction factor, unity for plain surfaces, but less for corrugated surfaces; θ is the rise of the surface temperature above the ambient air temperature, °C; and n has a value of 1.0 to 1.25, depending on the shape and position of the surface. For plain vertical surfaces about 2 ft and upward, in air, for temperature rises up to 70°C, at sea level, $n = 1.25$. For air densities at various elevations, W_c varies as the square root of the air density.

A comparison of Eq. (12-14) may be made with the familiar expression for surface conductance for free convection over large vertical surfaces in air, in the form $h_c = C_1 k (a\,\Delta t)^{1/3}$ [Eq. (8-11)]. Since the heat transfer per square foot is the product of Δt and h_c, an expression for heat transfer, q/A, based on Eq. (8-11), would be $q/A = \Delta t\, C_1 k (a\,\Delta t)^{1/3}$, or $C_1 k a^{1/3} \Delta t^{4/3}$. Numerical results found by the two expressions show fair agreement.

The following example will illustrate the application of the principles of heat transfer to the design of a transformer.

Example 12-7. A 13,200-volt, single-phase, 60-cycle transformer of the oil-immersed self-cooled type has windings with an external surface of 31.6 sq ft. The outside dimensions of the tank are 41 by 25 in. by a height of

91 in. Tests of a transformer of similar design show the temperature rise of the hot spot over the adjacent* oil, θ, in degrees centigrade, to be 18.25 $L^{0.75}\mu^{0.25}$, where L is the loss in watts per square inch of winding surface and μ is the viscosity of the oil in centipoises at the film temperature.

The capacity of this transformer is to be increased by the addition of steel tubes of 1.03 in. i.d. and 1.25 in. o.d. set vertically on all four sides of the tank attached near the top and bottom and averaging 70 in. in length. The radiating enveloping surface of the tank and tubes will be 153 sq ft. The desired capacity is 500 kva, with an anticipated loss of 6,000 watts. The hot-spot temperature of the windings is not to exceed 95°C when the ambient air temperature is 30°C. The transformer is to be in a shaded location and is to be painted with paint which has an emissivity of 0.92. Determine the required number of tubes for the increased cooling of the oil.

Solution. In order to evaluate the physical properties of the oil and air at the film temperature it is necessary to estimate the various temperatures. This can be done, for a first approximation, by comparing the resistances which the heat flow will encounter. The four resistances may be expressed in sequence, as $1/A_1h_1$, $1/A_2h_2$, x/A_3k_m, and $1/A_4h_4$, where A_1 is the surface of the windings, A_2 is the inside surface of the tanks and tubes, A_3 is the mean surface of the tank and tubes, x is the thickness and k_m is the thermal conductivity of the metal, A_4 is the outside surface of the tank and tubes, and h_1, h_2, and h_4 are the surface conductances. Not all these values can be estimated with any satisfactory degree of accuracy, but it may be judged that the greatest resistance, and consequently the greatest temperature drop, will occur at the outside air film, even though that surface will be the greatest. The resistance of the oil film on the surface of the windings will no doubt be considerably greater than the resistance of the oil film on the inside surface of the tank and tubes, because of the great difference in extent of surfaces. The resistance of the metal will no doubt be the smallest of the four resistances. From these observations it would appear that the mean temperature of the oil may be estimated, for a first approximation, as the mean of the hot-spot temperature of the windings and the temperature of the ambient air. The temperature of the oil would then be $(95 + 30)/2 = 62.5°C$. The temperature of the oil adjacent to (1.5 in. from) the hot spot may be assumed to be $62.5 + 2 = 64.5°C$, and the rise of the hot spot over the adjacent oil temperature will be 30.5°C. The temperature of the oil film on the surface of the windings then is $64.5 + (\frac{3}{8} \times 30.5) = 75.9°C$. Evaluation of the viscosity of the oil at 75.9° (Table 12-2) and its substitution into Eq. (12-12) in the form $\theta = 18.25 \times [6000/(144 \times 31.6)]^{0.75} \times 3.054^{0.25} = 29.7°C$. The temperature of the adjacent oil will then be $95 - 29.7 = 65.3°C$, and the mean temperature of the main body of oil in the tank will be $65.3 - 2 = 63.3°C$. A second approximation based upon an assumed temperature of the adjacent oil of 65.3° yields the value of θ as 29.6°C and an adjacent oil temperature of 65.4°C. The average temperature of the oil will then be $65.4 - 2 = 63.4°C$. Since the average

---

* Measured 1.5 in. from coil insulation surface.

temperature of the oil is 33.4°C above the ambient air temperature of 30°C, some 30°C of the temperature drop may be assumed to occur at the outer or air-side surface. On this basis the transfer by radiation and free convection may be computed for the assumed surface temperature of 30 + 30 = 60°C. By Eq. (12-13) the heat transfer by radiation is $W_r = K\epsilon(T_1^4 - T_2^4) =$ 3.68 × 10^{-11} × $0.92(333^4 - 303^4) = 0.1302$ watt/sq in., or for an enveloping surface of 153 sq ft, $q_r = 0.1302 \times 153 \times 144 = 2,875$ watts. By Eq. (12-14) the heat transfer by free convection of air for an empirically established friction factor $F = 0.95$ is $W_c = 0.0014 F \theta^n = 0.0014 \times 0.95 \times 30^{1.25}$ = 0.0933 watt/sq in. The heat to be removed by free convection equals 6,000 − 2,875 = 3,125 watts, and the required outside surface will be 3125/(0.0933 × 144) = 233 sq ft. The corresponding inside surface will be approximately 210 sq ft, and the mean surface approximately 220 sq ft. These values are tentative and must be verified by establishing the complete temperature gradient. The resistance of the metal (assuming a flat surface having an average thickness of 0.15 in.) will be $x/A_3 k_m =$ 0.15/(12 × 220 × 34) = 0.000,001,67. By comparison, the resistance of the oil on the surface of the windings at hot-spot temperature is $1/A_1 h_1$ where $A_1 = 31.6$ sq ft. $h_1 = (6000 \times 3.413)/(29.6°C \times \frac{9}{5} \times 31.6) =$ 12.17 Btu/(hr)(sq ft)(°F).

$$\frac{1}{A_1 h_1} = \frac{1}{31.6 \times 12.17} = 0.002605$$

The temperature drop through the metal will be (0.000,001,67/0.002605) × 29.6 = 0.019°C—too small a value for any further consideration. The inner-surface temperature of the tank and tubes will be 60°C. The mean temperature of the oil film on the inside surface will be (60 + 63.4)/2 = 61.7°C (or 143.1°F), and the estimated temperature drop across the film will be 63.4 − 60.0 = 3.4°C (or 6.12°F).

$$aL^3\, \Delta t = 3.8 \times 10^8 \times 8 \times 6.12 = 18.6 \times 10^9$$

By Eq. (8-10)

$$h_c = C_1 \frac{k}{L} (aL^3\, \Delta t)^{1/3}$$

$$= 0.13 \times \frac{0.078}{2} (18.6 \times 10^9)^{1/3} = 13.42 \text{ Btu/(hr)(sq ft)(°F)}$$

$q = A h_c\, \Delta t = 210 \times 13.42 \times 6.12 = 17,250$ Btu/hr. But for a desired transfer of 6,000 watts, or 20,478 Btu, an increase in $A h_c\, \Delta t$ must be made—mainly by an increase in Δt. For a revised value of $\Delta t = 3.9°C$ (or 7.02°F)

$$aL^3\, \Delta t = 21.3 \times 10^9$$
$$h_c = 0.13 \times \frac{0.078}{2} (21.3 \times 10^9)^{1/3} = 14.06 \text{ Btu/(hr)(sq ft)(°F)}$$

$q = 210 \times 14.06 \times 7.02 = 20,650$ Btu/hr, which is close to the desired transfer of 6,000 watts, or 20,478 Btu. On the basis of a temperature

drop of 3.9°C across the oil film, the outside-surface temperature will be
63.4 − 3.9 = 59.5°C. A revised calculation of the heat transfer from the
outer surface by radiation and convection will yield

$$W_r = 3.68 \times 10^{-11} \times 0.92(332.5^4 - 303^4) = 0.1273 \text{ watt/sq in.}$$

$q_r = 0.1273 \times 153 \times 144 = 2{,}805$ watts. By Eq. (12-14)

$$W_c = 0.0014 F \theta^n$$
$$= 0.0014 \times 0.95 \times 29.5^{1.25} = 0.915 \text{ watt/sq in}$$

The heat to be removed by free convection = 6,000 − 2,805 = 3,195 watts,
and the required outside surface will be 3,195/(0.915 × 144) = 242 sq ft.
The corresponding inside surface will be approximately 216 sq ft, and the
mean surface approximately 229 sq ft. Recalculation, based upon the
revised estimates of the surfaces, yields the following values: for the temperature drops, °C,

Hot spot to adjacent oil.............	29.6
Adjacent oil to average oil..........	2.0
Average oil to metal................	3.8
	35.4

The temperature of the outside surface will then be 95 − 35.4 = 59.6°C,
and the heat transfer from the outer surface by radiation and free convection
will be as follows:

$$W_r = 3.68 \times 10^{-11} \times 0.92(332.6^4 - 303^4) = 0.1286 \text{ watt/sq in.}$$
$$q_r = 0.1286 \times 153 \times 144 = 2{,}835 \text{ watts}$$
$$W_c = 0.0014 \times 0.95 \times 29.6^{1.25} = 0.0918 \text{ watt/sq in.}$$

The heat to be removed by free convection = 6,000 − 2,835 = 3,165 watts;
and the required outside surface = 3,165/(0.0918 × 144) = 239 sq ft. The
required tube surface beyond that of the tank = 239 − 97.6 = 141.4 sq ft.
The required number of tubes of 70-in. length = 141.4 × 144/(70 × 1.25π)
= 74 tubes.

COMMENTS: The values of temperature drop across the two oil films show a
great difference—29.6°C from the hot spot to the adjacent oil and only
3.8°C across the oil·film on the inner surfaces of the tank and tubes. It
should be observed, however, that only about three-fourths of the 29.6° drop,
or 22.2°, occurs across the oil film; the remainder is across the insulation on
the windings. In addition, because of the difference in the surface of the
tank and tubes and the surface of the windings, the drop through the oil
film on the windings might be expected to be approximately 6.8 times the
drop through the oil film on the inside of the tank and tubes for the same
surface conductance. For a drop of 3.8°C across the oil film on the inside
of the tank and tubes the corresponding drop across the oil film on the windings would be 6.8 × 3.8, or 25.8°C. This value appears to be in reasonably
close agreement with the estimated value of 22.2°C, especially since in the
comparison no account has been taken of the difference in the conductance
of the two oil films at the different temperatures or of the variation in the

temperature of the different parts of the windings. The large difference between the temperature drops across the oil films on the windings and on the inner surface of the tank and tubes apparently is to be expected.

Problems

1. Design a closed heater for heating the feed water to be supplied to a 250-hp boiler which may be operated at 175 per cent of rating, delivering steam at 180 psig and a quality of 99 per cent. The water is to be heated from 100 to 190°F, and the necessary heat is to be furnished by saturated steam at 225°F. Because of space limitations a heater of the vertical two-pass type, with tubes not more than 12 ft in length, is desired.

2. Design a two-pass surface condenser to condense 48,000 lb of exhaust steam per hour at a temperature of 94°F. An adequate but not a liberal supply of slightly muddy cooling water at a maximum temperature of 74°F is available.

3. A fin-tube fan coil is to be designed to heat 9,000 cu ft of air per minute from 20 to 90°F. The heating medium is to be saturated steam at 215°F. Tubing is to be $\frac{1}{2}$ in. i.d. and $\frac{5}{8}$ in. o.d., and the ratio of the outer surface of the fins and tubes to the inner surface of the tubes is to be 20:1. The net air opening is to be 50 per cent of the face area. Tests of a coil of similar type show the outside-surface coefficient h_o to be represented by the expression

$$h_o = 0.047 \frac{G^{0.52}}{D^{0.48}} \quad \text{Btu/(hr)(sq ft)(°F)}$$

where G = mass velocity in $\text{lb}_m/(\text{hr})(\text{sq ft of net air opening})$ and D = outside diameter of the tube in feet.

Design the coil for a face velocity of 550 fpm. Determine the face area and the area of the external surface of the coil.

4. What will be the exit-air conditions, the sensible-, latent-, and total-heat transfer, and the refrigerant temperature for the coil described in Ex. 12-5 if the cooling medium is dichlorodifluoromethane (Freon-12)?

5. A fin-tube coil is to be designed to cool air at an initial condition of 100°F dry-bulb and 79°F wet-bulb to a final condition of 65°F dry-bulb and 61°F wet-bulb temperature. The outside-surface area of the type of coil selected is 18 sq ft/(sq ft of face area)(row of coil depth). If the inlet air velocity over the face area is 350 fpm, at which tests of a similar coil show an outside-surface coefficient of 10.0, determine the required number of rows of coil depth.

6. Determine the refrigerant temperature required to produce a coil surface temperature of 48°F when a coil, which has four rows of tubes in depth, delivers 1.0 ton of refrigeration per square foot of face area. For each square foot of face area the coil has a total external surface of 224 sq ft and an internal surface of 18 sq ft. Assume a conductance for the coil and its inside surface of 300 Btu/(hr)(sq ft)(°F).

7. Determine the required number of tubes to be connected to the electrical transformer described in Ex. 12-7 if the external surface of the windings is increased from 31.6 to 35.0 sq ft.

REFERENCES

1. Schack, A., H. Goldschmidt, and E. P. Partridge; "Industrial Heat Transfer," John Wiley & Sons, Inc., New York, 1933.
2. Drew, T. B., H. C. Hottel, and W. H. McAdams: *Trans. Am. Inst. Chem. Engrs.*, **32**, 271–305 (1936).
3. Standards of Tubular Exchanger Manufacturers Association, pp. 49–51, New York, 1949.
4. Gaffert, G. A., "Steam Power Stations," pp. 107–142, McGraw-Hill Book Company, Inc., New York, 1952.
5. Orrok, G. A., Transmission of Heat in Surface Condensers, *Trans. ASME*, **32**, 1139 (1910).
6. "Standards for Steam Condensers," Heat Exchange Institute, New York, 1955.
7. Ferguson, R. M., and J. C. Oakden: Heat Transfer Coefficients for Water and Steam in a Surface Condenser, *Chem. Eng. Congr., World Power Conf., London, No.* 114, 1936.
8. Van Hengel, H.: Tests of a 50,000 Sq. Ft. Surface Condenser at Widely Varying Temperatures, Velocities of Inlet Water, and Loads, *Trans. ASME*, **58**, 627–641 (1936).
9. Goodman, W.: Dehumidification of Air with Coils, *Refrig. Eng.*, **32**(4), 225 (October, 1936).
10. Tuve, G. L., and L. J. Siegel: Performance of Surface-coil Dehumidifiers, *Trans. ASHVE*, **44**, 523–548 (1938).
11. Tuve, G. L., and C. A. McKeeman: Heat Transfer from Direct and Extended Surfaces with Forced Air Circulation, *Trans. ASHVE*, **40**, 427–442 (1934).
12. Air Heating and Cooling Coils, chap. 36, pp. 829–846, in *Heating, Ventilating, Air Conditioning Guide*, American Society of Heating and Air-conditioning Engineers, New York, 1956.
13. Colburn, A. P., and O. A. Hougen, Design of Cooler Condensers for Mixtures of Vapors with Noncondensing Gases, *Ind. Eng. Chem.*, **26**, 1178–1182 (1934).
14. Apparatus Dew Point, chap. 13, pp. 320–322, in *Heating, Ventilating, Air Conditioning Guide*, American Society of Heating and Air-conditioning Engineers, New York, 1956.
15. Performance of Dehumidifying Coils, chap. 36, pp. 839–846, in *Heating, Ventilating Guide*, American Society of Heating and Air-conditioning Engineers, New York, 1955.
16. Knowlton, A. E. (ed.): "Standard Handbook for Electrical Engineers," pp. 547–644, McGraw-Hill Book Company, Inc., New York, 1949.

CHAPTER 13

TRANSIENT CONDUCTION

13-1. General. In many practical problems it is necessary to determine the heat flow, the temperature distribution, and the time required for any point within a solid body to attain a given temperature when the body, initially at a uniform temperature, is subjected to a higher or lower surface temperature. Such a problem is encountered in the mold curing of rubber and other plastics, where the temperature of the surface of the material may be assumed to be changed from room temperature to the surface temperature of the mold. Somewhat different conditions are imposed in the case of heating metals in a heat-treating furnace and the heating of lumber in kilns. In this case the altered surface temperature cannot be assumed to be the same as that of the surrounding medium, but must be different, depending upon the surface coefficient. A still different case is one in which the surface of a solid, initially at a constant uniform temperature, is subjected to a variable-temperature fluid. This is the condition met in problems dealing with the temperatures attained by pistons and cylinder walls of steam engines and internal-combustion engines and in a regenerator.

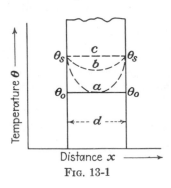

Fig. 13-1

All the examples cited are cases of variable heat flow in which the temperature distribution within the body, instead of remaining constant, changes with time.

Assume that the wall of thickness d in Fig. 13-1 is originally at a uniform temperature θ_o and is suddenly subjected to a higher surface temperature θ_s on both surfaces. At the outset, before the surface temperature is changed, the temperature at any

point in the body is represented by the line $\theta_o a \theta_o$. After the
temperature of both surfaces is changed to θ_s, the temperature
distribution may be represented by the line $\theta_s a \theta_s$; some time
later the temperature distribution may be represented by the
line $\theta_s b \theta_s$; and finally, after sufficient time has passed, the temperature will again be uniform and will be given by the straight
line $\theta_s c \theta_s$. In this case heat flows into the wall from both sides
and is stored in the wall, with a subsequent increase in temperature as time passes. As the temperature at any point in the body
increases, the rate of heat flow decreases until the new steady
state is reached.

The temperature distribution in a body for the unsteady-
state condition is given by Fourier's general law of heat conduction, which is a partial differential equation derived by the usual
methods. It is

$$\frac{\partial \theta}{\partial t} = \frac{k}{\rho c_p}\left(\frac{\partial^2 \theta}{\partial x^2} + \frac{\partial^2 \theta}{\partial y^2} + \frac{\partial^2 \theta}{\partial z^2}\right) \tag{13-1}$$

where θ = temperature at any point given by the coordinates
$\quad\quad x,\ y,\ z$.
$\quad t$ = time.
$\quad k$ = coefficient of thermal conductivity.
$\quad c_p$ = specific heat of unit mass.
$\quad \rho$ = mass per unit volume.

There are various solutions of Eq. (13-1) which may be found
by applying the standard methods for the solution of partial
differential equations. Any good text on advanced calculus or
on boundary-value problems will describe several particular
solutions. It is necessary to select the proper solution so that
the initial and boundary conditions of the problem may be satisfied. For analytical solutions to a number of specific problems
the reader is referred to several of the excellent texts on heat
conduction.[1,2,3]

In all problems of unsteady-state heat flow it is found that
the term $k/\rho c_p$ is involved. From Fourier's equation, for example, it is seen that the rate of temperature change with change in
time is dependent upon this term. In a practical case this simply
means that the rate of temperature change with change in time is
more rapid in a body that combines a high coefficient of thermal
conductivity with a low specific heat per unit volume. For
example, the temperature at the center of a slab of metal that

is subjected to a sudden increase in surface temperature would increase much more rapidly than the temperature of an exactly similar slab of wood subjected to the same temperature conditions. Thus a body having a high value of $k/\rho c_p$ heats faster than a body having a low $k/\rho c_p$ value. Since this term is always involved in unsteady-state heat flow, it is given a name, *thermal diffusivity*, and it will be designated by the letter α. In the dimension system used in this text the dimensions of α are L^2/T, and the units of α are square feet per hour.

In the following sections will be treated a number of practical problems that have solutions of simple form to be obtained by the methods of dimensional analysis. As in all analysis by the dimensional method, the value of the dimensionless functions must be obtained by experiment or by some other type of analysis. In the following solutions the values of the function used are given graphically and may be obtained from the solutions of Fourier's equation.

13-2. Infinitely Thick Wall Initially at a Constant Uniform Temperature θ_o, Subsequently Subjected to a Single Sudden Change in the Temperature of the Surface to θ_s. In Fig. 13-2 a portion of an infinitely thick wall is shown. The temperature curve for the initial state is indicated by the broken line $\theta_o\theta_o$. If the temperature of the surface is suddenly altered to θ_s, then at some time t the temperature at any plane parallel to the surface and a distance x away from it is θ, and the temperature curve is given by the solid line $\theta_s\theta$. The temperature difference $\theta_s - \theta$ is dependent upon the change in surface temperature $\theta_s - \theta_o$, the distance x, the time t, and the thermal diffusivity α. The variables involved and the dimensions are as shown in the accompanying table.

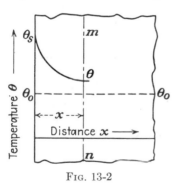

FIG. 13-2

Applying the π theorem to these variables it is seen that

$$\pi = \phi[(\theta_s - \theta)^e(\theta_s - \theta_o)^f x^g \alpha^h t^m]$$

or

$$\pi = \theta^e \theta^f L^g (L^2/T)^h T^m$$

hence

$$\pi = \theta^{e+f} L^{g+2h} T^{m-h}$$

Variable	Symbol	Dimensional formulas
Temperature difference between surface and plane mn at time t	$\theta_s - \theta$	θ
Change in surface temperature	$\theta_s - \theta_o$	θ
Distance from surface to plane mn	x	L
Thermal diffusivity of material	α	L^2/T
Time during which change in surface temperature has existed	t	T

Equating the exponents to zero yields the equations

$$e + f = 0$$
$$g + 2h = 0$$
$$m - h = 0$$

For π_1, let $e = 1$ and $m = 0$; then $f = -1$, $g = 0$, and $h = 0$. Therefore,

$$\pi_1 = \frac{\theta_s - \theta}{\theta_s - \theta_o}$$

For π_2, let $e = 0$ and $g = 1$; then $f = 0$, $h = -\frac{1}{2}$, and $m = -\frac{1}{2}$, which gives

$$\pi_2 = \frac{x}{\sqrt{\alpha t}}$$

The equation $\psi(\pi_1, \pi_2) = 0$, as indicated by the π theorem, then becomes

$$\psi \left(\frac{\theta_s - \theta}{\theta_s - \theta_o}, \frac{x}{\sqrt{\alpha t}} \right) = 0$$

or

$$\frac{\theta_s - \theta}{\theta_s - \theta_o} = f_1 \left(\frac{x}{\sqrt{\alpha t}} \right)$$

From this expression an equation for the temperature θ at any plane mn and at any time t is obtained. This is

$$\theta = \theta_s - (\theta_s - \theta_o)f_1 \left(\frac{x}{\sqrt{\alpha t}} \right) \quad °F \qquad (13\text{-}2)$$

Values of $f_1(x/\sqrt{\alpha t})$ for various values of $x/\sqrt{\alpha t}$ are given in the curve of Fig. 13-3. Although Eq. (13-2) has been derived for a wall of infinite thickness, it may be used also for a wall of

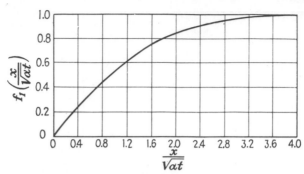

FIG. 13-3. Value of function f_1 for use with Eq. (13-2).

finite thickness if the temperature on only one side of the wall is changed to θ_s and if $d > 1.2\sqrt{\alpha t}$ (d equals wall thickness in feet). This approximation for a finite wall holds true only if its initial temperature is low and approximately the same as its surroundings.

The rate of heat flow q/A past any plane mn (Fig. 13-2) may be assumed to depend upon the change in surface temperature $\theta_s - \theta_o$, the distance x from the surface to the plane, the time t during which the temperature change has existed, the thermal diffusivity α, and the coefficient of thermal conductivity k. The dimensions of these variables are as follows:

Variable	Dimensional formulas
q/A	H/L^2T
$\theta_s - \theta_o$	θ
x	L
α	L^2/T
t	T
k	$H/LT\theta$

The general equation for π is

$$\pi = \phi[(q/A)^e(\theta_s - \theta_o)^f x^g \alpha^h t^m k^n]$$

and substitution of the dimensions yields the equation

$$\pi = H^{e+n}\theta^{f-n}L^{-2e+g+2h-n}T^{-e-h+m-n}$$

for which the simultaneous equations

$$e + n = 0$$
$$f - n = 0$$
$$-2e + g + 2h - n = 0$$
$$-e - h + m - n = 0$$

must be satisfied.   For π_1, let $e = 1$ and $g = 0$; then $f = -1$, $h = \frac{1}{2}$, $m = \frac{1}{2}$, $n = -1$, and

$$\pi_1 = \frac{q\sqrt{\alpha t}}{A(\theta_s - \theta_o)k}$$

Similarly for π_2, let $e = 0$ and $h = -1$; then $f = 0$, $g = 2$, $h = -1$, $m = -1$, $n = 0$, and

$$\pi_2 = \frac{x^2}{\alpha t}$$

Then

$$\psi(\pi_1, \pi_2) = 0$$

becomes

$$\psi\left[\frac{q\sqrt{\alpha t}}{A(\theta_s - \theta_o)k}, \frac{x^2}{\alpha t}\right] = 0$$

which has a solution

$$\frac{q\sqrt{\alpha t}}{A(\theta_s - \theta_o)k} = f_2\left(\frac{x^2}{\alpha t}\right)$$

or $$\frac{q}{A} = \frac{k(\theta_s - \theta_o)}{\sqrt{\alpha t}} f_2\left(\frac{x^2}{\alpha t}\right) \qquad \text{Btu/(sq ft)(hr)} \qquad (13\text{-}3)$$

where q/A = rate of heat flow past any plane mn after t hr. Values of $f_2(x^2/\alpha t)$ for various values of $x^2/\alpha t$ are given in the curve of Fig. 13-4.   From this curve it is seen that the rate of heat flow through the surface $x = 0$ at the end of t hr is

$$\frac{q_s}{A} = \frac{0.565k(\theta_s - \theta_o)}{\sqrt{\alpha t}} \qquad \text{Btu/(sq ft)(hr)} \qquad (13\text{-}4)$$

The total heat flow past the surface in t hr is obtained by integrating Eq. (13-4) over the time interval $t = 0$ to $t = t$.   The total heat flow is

$$\frac{Q}{A} = 1.13k(\theta_s - \theta_o)\sqrt{\frac{t}{\alpha}} \qquad \text{Btu/sq ft} \qquad (13\text{-}5)$$

Equations (13-3), (13-4), and (13-5) are used chiefly in determining the heat flow in very thick walls, i.e., where the wall thickness is greater than $1.2\sqrt{\alpha t}$ and where the initial temperature of the wall is low and approximately the same as its surroundings.

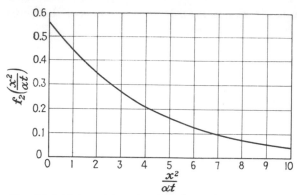

FIG. 13-4. Value of function f_2 for use with Eq. (13-3).

Example 13-1. The temperature of the entire surface of one side of a steel slab 10 in. thick is suddenly changed from 60 to 300°F. Find the temperature and the rate of heat flow 3 in. below this surface after 12 min has elapsed. Find also the rate of heat flow through the surface at the end of 12 min, and the total heat flow through the surface during the 12 min.

Solution. The specific weight of steel is approximately 490 lb/cu ft; its mean specific heat is 0.11; and its thermal conductivity is 34 Btu/(hr) (sq ft)(°F/ft). Then

$$\alpha = \frac{34}{0.11 \times 490} = 0.631 \text{ sq ft/hr}$$
$$x = \tfrac{3}{12} = 0.25 \text{ ft}$$
$$d = \tfrac{10}{12} = 0.833 \text{ ft}$$

and
$$t = \tfrac{12}{60} = 0.2 \text{ hr}$$

Now

$$1.2 \sqrt{\alpha t} = 1.2 \sqrt{0.631 \times 0.2} = 0.426$$

and since this is less than the thickness d, Eq. (13-2) may be used. Thus

$$\theta = \theta_s - (\theta_s - \theta_o)f_1 \left(\frac{x}{\sqrt{\alpha t}} \right)$$
$$= 300 - (300 - 60)f_1 \left(\frac{0.25}{\sqrt{0.631 \times 0.2}} \right)$$
$$= 300 - 240f_1(0.705)$$

From Fig. 13-3 the value of $f_1(0.705)$ is found to be 0.37, hence

$$\theta = 300 - (240 \times 0.37)$$
$$= 211.2°F$$

The rate of heat flow through the plane 3 in. below the heated surface after 12 min. is found from Eq. (13-3). This is

$$\frac{q}{A} = \frac{k(\theta_s - \theta_o)}{\sqrt{\alpha t}} f_2 \left(\frac{x^2}{\alpha t}\right)$$

$$= \frac{34(300 - 60)f_2}{\sqrt{0.631 \times 0.2}} \left(\frac{0.25^2}{0.631 \times 0.2}\right)$$

$$= 22{,}970 f_2(0.495)$$

The value $f_2(0.495) = 0.5$ is obtained from Fig. 13-4. q/A, therefore, is equal to $22{,}970 \times 0.5 = 11{,}485$ Btu/(hr)(sq ft).

The rate of heat flow through the surface at the end of 12 min is determined by substitution in Eq. (13-4), which is

$$\frac{q_s}{A} = \frac{0.565k(\theta_s - \theta_o)}{\sqrt{\alpha t}}$$

$$= \frac{0.565 \times 34(300 - 60)}{\sqrt{0.631 \times 0.2}}$$

$$= 12{,}980 \text{ Btu/(sq ft)(hr)}$$

The total heat flow through the surface during the 12 min is given by Eq. (13-5). This is

$$\frac{Q}{A} = 1.13k(\theta_s - \theta_o) \sqrt{\frac{t}{\alpha}}$$

$$= 1.13 \times 34(300 - 60) \sqrt{\frac{0.2}{0.631}}$$

$$= 5193 \text{ Btu/sq ft}$$

13-3. Walls of Finite Thickness Initially at a Uniform Temperature θ_o, Subsequently Subjected to a Single Sudden Change in the Temperature of Both Surfaces to θ_s. In this case the wall of Fig. 13-5 is assumed to have infinite lateral extension but a finite thickness d. Since the wall, in this case, is subjected to a change in surface temperature on both sides, then it must be true that the difference between the altered surface temperature θ_s and the temperature θ of the plane mn must be dependent upon the thickness of the wall as well as the other variables given for the infinitely thick wall. The variables and their dimensions are as follows:

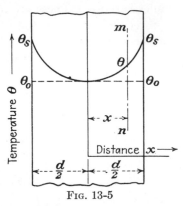

FIG. 13-5

Variable	Symbol	Dimensional formulas
Temperature difference between surface and plane mn at time t...........................	$\theta_s - \theta$	θ
Change in surface temperature.................	$\theta_s - \theta_o$	θ
Distance from center to plane mn...............	x	L
Thickness of wall.............................	d	L
Thermal diffusivity of the material..............	α	L^2/T
Time during which change in surface temperature has existed................................	t	T

With these variables, the equation for π is

$$\pi = \phi[(\theta_s - \theta)^e(\theta_s - \theta_o)^f x^g d^h \alpha^m t^n]$$

and in terms of the dimensions

$$\pi = \theta^{e+f} L^{g+h+2m} T^{-m+n}$$

Equating the exponents to zero:

$$e + f = 0$$
$$g + h + 2m = 0$$
$$-m + n = 0$$

In this case there are six variables expressed in terms of three fundamental dimensions. There will be, therefore, three dimensionless groups. These are found by assigning arbitrary values to three unknowns in the simultaneous equations and solving for the other three. Thus if $e = 1$, $g = 0$, and $m = 0$, then $f = -1$, $h = 0$, and $n = 0$. These values yield the equation $\pi_1 = (\theta_s - \theta)/(\theta_s - \theta_o)$. Similarly if $e = 0$, $m = 1$, and $g = 0$, then $f = 0$, $h = -2$, and $n = 1$. This solution gives the equation $\pi_2 = \alpha t/d^2$. For the last solution let $e = 0$, $g = 1$, and $m = 0$; then $f = 0$, $h = -1$, and $n = 0$. When these values are substituted in the expression for π, the result is $\pi_3 = x/d$. The general solution then is

$$\psi(\pi_1, \pi_2, \pi_3) = 0$$

or
$$\psi\left(\frac{\theta_s - \theta}{\theta_s - \theta_o}, \frac{\alpha t}{d^2}, \frac{x}{d}\right) = 0$$

A particular solution for this equation is

$$\frac{\theta_s - \theta}{\theta_s - \theta_o} = f_3\left(\frac{\alpha t}{d^2}, \frac{x}{d}\right)$$

or $$\qquad \theta = \theta_s - (\theta_s - \theta_o)f_3\left(\frac{\alpha t}{d^2}, \frac{x}{d}\right) \qquad °F \qquad (13\text{-}6)$$

The values of $f_3(\alpha t/d^2, x/d)$ are given in the curves of Fig. 13-6.

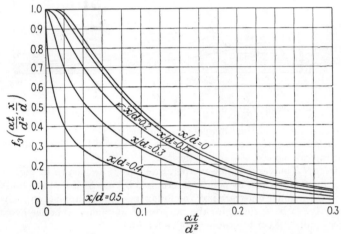

FIG. 13-6. Value of function f_3 for use with Eq. (13-6).

An equation for the rate of heat flow past any plane mn may be found by the method that was used in the case of the infinitely thick wall.*

In this case it is found that

$$\frac{q}{A} = (\theta_s - \theta_o)\frac{k}{d}f_4\left(\frac{\alpha t}{d^2}, \frac{x}{d}\right) \qquad Btu/(hr)(sq\ ft) \qquad (13\text{-}7)$$

Values of $f_4(\alpha t/d^2, x/d)$ are given in the curves of Fig. 13-7.

Example 13-2.   A slab of rubber, 1 in. thick, initially at a temperature of 80°F, is cured in a mold having surface temperatures of 350°F.   It is desired

---

* The student may derive this equation.   It must be remembered that in the case of a finite wall the rate of heat flow is dependent upon the temperature change of the wall surface, the distance x from the mid-plane to the plane mn, the thermal diffusivity α, the time t, the coefficient of thermal conductivity k, and the thickness of the wall d.

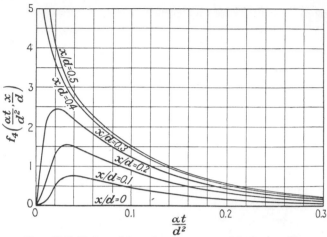

FIG. 13-7. Value of function f_4 for use with Eq. (13-7).

to determine a set of time-temperature curves for points at $\frac{1}{8}$, $\frac{1}{4}$, $\frac{3}{8}$, and $\frac{1}{2}$ in. below the surface of the slab. Also, the rate of heat flow $\frac{1}{2}$ in. below the surface at the end of 3 min is desired.

Solution. The desired time-temperature curves are shown in Fig. 13-7a. These curves are a plot of points determined by use of Eq. (13-6) and Fig. 13-6, in the manner illustrated by the following solution applied to the point marked A. It is assumed, for the purpose of simplicity, that the average* physical properties of the rubber are as follows:

$$k = 0.1 \text{ Btu/(hr)(sq ft)(°F/ft)}$$
$$\rho = 94 \text{ lb}_m/\text{cu ft}$$
$$c_p = 0.40 \text{ Btu/(lb}_m)(°F)$$

Since this point is $\frac{1}{8}$ in. below the surface, $x = \frac{3}{8} \div 12 = 0.03125$ ft (see Fig. 13-5), and $d = \frac{1}{12} = 0.0833$ ft. Thus

$$\alpha = 0.1/(0.4 \times 94) = 0.00266 \text{ sq ft/hr}$$

$$t = 30 \text{ min} = 0.5 \text{ hr}$$

$$\alpha t/d^2 = (0.00266 \times 0.5)/0.0833^2 = 0.19$$

$$x/d = 0.03125/0.0833 = 0.375$$

When these values are substituted in Eq. (13-6), the equation

$$\theta = 350 - (350 - 80)f_3(0.19, 0.375)$$

is obtained. From Fig. 13-6, $f_3(0.19, 0.375) = 0.08$, and therefore

$$\theta = 350 - (270 \times 0.08)$$
$$\theta = 328°F$$

---

* This simplifying assumption is necessary, since the properties of rubber will change somewhat, owing to the chemical change which takes place.

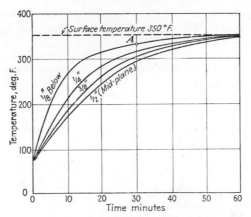

Fig. 13-7a. Heating curves for 1-in. rubber slab.

The rate of heat flow past any plane at any time may be found by the use of Eq. (13-7). Thus the rate of heat flow past the surface $x/d = 0.5$ at the end of 3 min ($t = 0.05$ hr) is

$$\frac{q}{A} = (350 - 80) \frac{0.1}{0.0833} f_4 \left(\frac{0.00266 \times 0.05}{0.0833^2}, 0.5 \right)$$
$$= 324.3 f_4 (0.019, 0.5)$$

From Fig. 13-7, $f_4(0.019, 0.5) = 4.1$. Therefore

$$\frac{q}{A} = 324.3 \times 4.1$$
$$= 1330 \ \text{Btu}/(\text{hr})(\text{sq ft})$$

13-4. Temperature at the Centers of Various Bodies Initially at a Uniform Temperature θ_o, Subsequently Subjected to a Single Sudden Change in the Temperature of the Surface to θ_s. The temperature at any point within a sphere, a cube, or a square beam is more difficult to calculate than for the two cases already discussed, since the heat flow in such cases is not unidirectional. If the temperature at the center of the body is desired, however, it may be found by the method used in deriving Eq. (13-6). In this case the distance from the center to the surface is simply $d/2$ and the variable x may be eliminated. Thus the equation for the temperature at the center of a body becomes

$$\theta_c = \theta_s - (\theta_s - \theta_o) f_5 \left(\frac{\alpha t}{d^2} \right) \qquad °\text{F} \qquad (13-8)$$

Values of f_5 have been worked out for various shapes by E. D. Williamson and L. H. Adams.[4]   These values are plotted in the curves of Fig. 13-8.

Example 13-3.   A 10-in. steel cube is quenched in water after it is removed from a heat-treating furnace that is maintained at a temperature of 1500°F. Determine the cooling curve for the center of the cube.

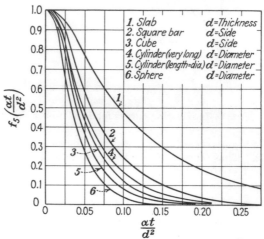

FIG. 13-8. Value of function f_5 for use with Eq. (13-8).   (*H. Groeber, "Einführung in die Lehre von der Wärmeübertragung," Springer-Verlag OHG, Vienna, 1926.*)

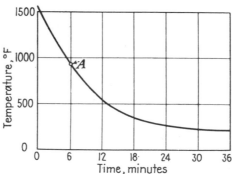

FIG. 13-8a. Cooling curve for 10-in. steel cube quenched in water.

Solution.   Equation (13-8) is applied in this case to determine the desired curve of Fig. (13-8a). θ_s may be assumed constant at 212°F for the purpose of this calculation, since the water adjacent to the surface would be boiling at atmospheric pressure.   The properties of steel at an average temperature of $(1500 + 212)/2 = 856$°F are $k = 22$ Btu/(hr)(sq ft)(°F/ft), $c_p = 0.17$ Btu/(lb$_m$)(°F), $\rho = 490$ lb$_m$/cu ft.   Thus for the conditions of this problem at the point A on the curve, $\alpha = 22/(0.17 \times 490) = 0.264$ sq ft/hr. $\theta_o = 1500°F$, $t = 6$ min $= 0.1$ hr, and $d = 10/12 = 0.833$ ft.*   Hence, according to Eq. (13-8)

$$\theta_c = 212 - (212 - 1500)f_5 \left(\frac{0.264 \times 0.1}{0.833^2}\right)$$
$$= 212 - (-1288)f_5(0.038)$$

From curve 3 of Fig. 13-8, the value $f_5(0.038) = 0.58$.   The temperature at the center, after 6 min, accordingly is

$$\theta_c = 212 + (1288 \times 0.58)$$
$$= 959°F$$

13-5. Wall of Finite Thickness Initially at a Uniform Temperature θ_o, Subsequently Exposed on Both Sides to a Liquid or Gas at a Constant Temperature θ_f.   In this case the temperature difference between the surface and the main body of the fluid, $\theta_s - \theta_f$, at any time is dependent upon the initial temperature difference between the body and the fluid, $\theta_o - \theta_f$; the thermal diffusivity α of the body; the coefficient of thermal conductivity k of the body; the thickness d of the body; the time t; and the surface coefficient h.†

Variables and their dimensions are as follows:

Variable	Dimensional formulas
$\theta_s - \theta_f$	θ
$\theta_o - \theta_f$	θ
α	L^2/T
k	$H/LT\theta$
d	L
t	T
h	$H/L^2T\theta.$

By applying the π theorem the following equations are obtained:

$$\pi = \psi[(\theta_s - \theta_f)^p(\theta_o - \theta_f)^r\alpha^s k^u d^v t^w h^z]$$

or

$$\pi = \theta^p\theta^r(L^2/T)^s(H/LT\theta)^u L^v T^w(H/L^2T\theta)^z$$

and

$$p + r - u - z = 0$$
$$2s - u + v - 2z = 0$$
$$-s - u + w - z = 0$$
$$u + z = 0$$

---

* d is the side of the cube, according to Fig. 13-8.

† h is the sum of the convection and radiation coefficients and is assumed to be constant.   The thermal diffusivity and the thermal conductivity are also assumed to be constant.

272   INTRODUCTION TO HEAT TRANSFER   [CHAP. 13

By making the proper substitutions the three π terms are found to be

$$\pi_1 = \frac{\theta_s - \theta_f}{\theta_o - \theta_f}$$

$$\pi_2 = \frac{\alpha t}{d^2}$$

$$\pi_3 = \frac{hd}{k}$$

from which the equation

$$\theta_s = \theta_f + (\theta_o - \theta_f)f_6\left(\frac{\alpha t}{d^2}, \frac{hd}{k}\right) \qquad °F \qquad (13\text{-}9)$$

is obtained.

Equation (13-9) may be used to determine the surface temperature θ_s of the wall at any time t after it has been exposed on

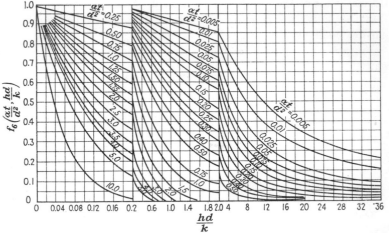

FIG. 13-9. Value of function f_6 for use with Eq. (13-9). (*H. Groeber, "Einführung in die Lehre von der Wärmeübertragung," Springer-Verlag OHG, Vienna, 1926.*)

both sides to a fluid having a temperature θ_f. Values of f_6 for various values of $\alpha t/d^2$ and hd/k are given in Fig. 13-9. The temperature at the center plane is given by the equation

$$\theta_m = \theta_f + (\theta_o - \theta_f)f_7\left(\frac{\alpha t}{d^2}, \frac{hd}{k}\right) \qquad °F \qquad (13\text{-}10)$$

which is derived in a manner similar to that used for deriving Eq. (13-9). In Eq. (13-10), θ_m is the temperature of the midplane of the slab and f_7 is a function the value of which may be found in Fig. 13-10.

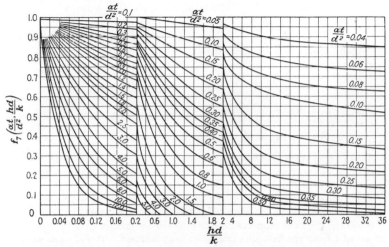

FIG. 13-10. Value of function f_7 for use with Eq. (13-10). (*H. Groeber, "Einführung in die Lehre von der Wärmeübertragung," Springer-Verlag OHG, Vienna, 1926.*)

Example 13-4. A steel plate 18 in. thick is to be cooled in a vertical position in air at 80°F from a temperature of 1500°F. It is required to determine the temperature of the surface and of the mid-plane of the plate at the end of 2 hr.

Solution. Equation (13-9) and Fig. 13-9 may be used to determine the surface temperature. First, however, it is necessary to decide upon the temperatures that will be used for determining the values of α, h, and k. Since the steel is initially at a temperature of 1500°F, and since its temperature is gradually decreasing, it seems logical to make a first assumption that the surface temperature is 1500°F and then later adjust the values to conform with the calculated temperature. Hence, for a first assumption, the properties of the steel will be determined at 1500°F and the value of h will be determined for an assumed surface temperature of 1500°F. Based on these assumptions, then, $k = 19.5$ Btu/(hr)(sq ft)(°F/ft), $c_p = 0.16$ Btu/(lb$_m$)(°F), $\rho = 490$ lb$_m$/cu ft,

$$\alpha = 19.5/(0.16 \times 490) = 0.249 \text{ sq ft/hr}$$

and $\alpha t/d^2 = (0.249 \times 2)/1.5^2 = 0.222$. The value of h may be found by Eq. (4-38) for h_r and by Eq. (8-9) for h_c. Thus

$$h_r = \frac{0.173 \times 10^{-8} \times 0.79(1960^4 - 540^4)}{1500 - 80}$$
$$= 14.1 \text{ Btu/(hr)(sq ft)(°F)}$$

and, for an average temperature of $(1500 + 80)/2 = 790$°F

$$h_c = 0.55 \times \frac{0.031}{2} [0.034 \times 10^6 \times 2^3(1500 - 80)]^{1/4}$$
$$= 1.2 \text{ Btu/(hr)(sq ft)(°F)}$$

Therefore, $h = h_r + h_c = 15.3$, and $hd/k = (15.3 \times 1.5)/19.5 = 1.18$. The surface temperature, according to Eq. (13-9), is

$$\theta_s = 80 + (1500 - 80)f_6(0.222, 1.18)$$

and since, from Fig. 13-9, $f_6(0.222, 1.18) = 0.55$, then

$$\theta_s = 80 + (1420 \times 0.55) = 860\text{°F}$$

This temperature is obviously low, since it was assumed that all the heat lost by radiation from the surface during the 2-hr period occurs at a temperature of 1500°F.

For a second approximation the average surface temperature during the 2-hr period may be assumed to be $(1500 + 860)/2 = 1180$°F.

Since the physical properties of the steel do not change greatly, the values of α and k from the first approximation may be used. Based upon an average surface temperature of 1180°F then,

$$h_r = \frac{0.173 \times 10^{-8} \times 0.79(1640^4 - 540^4)}{1180 - 80}$$
$$= 8.90 \text{ Btu/(hr)(sq ft)(°F)}$$

and for an average air temperature of $(1180 + 80)/2 = 630.0$°F

$$h_c = 0.55 \times \frac{0.0277}{2} [0.0622 \times 10^6 \times 2^3(1180 - 80)]^{1/4}$$
$$= 1.16 \text{ Btu/(hr)(sq ft)(°F)}$$

The surface coefficient is then

$$h = 8.90 + 1.16 = 10.06 \text{ Btu/(hr)(sq ft)(°F)}$$

and
$$\frac{hd}{k} = \frac{10.06 \times 1.5}{19.5} = 0.77$$

This new value may be substituted in Eq. (13-9), which becomes

$$\theta_s = 80 + (1500 - 80)f_6(0.222, 0.77)$$

and from Fig. 13-9

$$f_6(0.222, 0.77) = 0.67$$

hence

$$\theta_s = 80 + (1420 \times 0.67) = 1040\text{°F}$$

A third approximation, based on an average temperature of

$$\frac{1500 + 1040}{2} = 1270°F$$

shows that the surface temperature at the end of 2 hr is more nearly 990°F; and the result of a fourth approximation is very nearly 990°F.

The temperature of the center plane of the slab, at the end of 2 hr, may be found by assuming an average surface temperature of

$$\frac{1500 + 990}{2} = 1245°F$$

for finding the value of h to be substituted in Eq. (13-10). For this average surface temperature

$$h_r = \frac{0.173 \times 10^{-8} \times 0.79(1705^4 - 540^4)}{1245 - 80}$$
$$= 9.85 \text{ Btu/(hr)(sq ft)(°F)}$$

and for an average air temperature of $(1245 + 80)/2 = 663°F$

$$h_c = 0.55 \times \frac{0.0284}{2} [0.0536 \times 10^6 \times 2^3(1245 - 80)]^{1/4}$$
$$= 1.16 \text{ Btu/(hr)(sq ft)(°F)}$$

hence $\qquad h = 9.85 + 1.16 = 11.01 \text{ Btu/(hr)(sq ft)(°F)}$

therefore $\qquad \dfrac{hd}{k} = \dfrac{11.01 \times 1.5}{19.5} = 0.84$

When this value of hd/k and the value of $\alpha t/d^2$, which was determined in the first part of this example, are substituted in Eq. (13-10), it becomes

$$\theta_m = 80 + (1500 - 80)f_7(0.222, 0.84)$$

The value of $f_7(0.222, 0.84)$ in Fig. 13-10 is found to be 0.78; hence

$$\theta_m = 80 + (1420 \times 0.78)$$
$$= 1188°F$$

Problems

1. A steel cylinder, 6 in. in diameter and 6 in. long, is to be quenched in a water bath from a temperature of 1450°F. Determine the time required for the temperature at the center to drop to 290°F. Assume $c_p = 0.17$ and $\rho = 490 \text{ lb}_m$ per cu ft.

2. A plastic material in the form of a 2-in. slab is to be cured in a flat-plate press having surfaces at a temperature of 300°F. If the material is initially at 70°F, how long will it take to raise the temperature at 0.8 in. below the surface to 280°F? Assume the following average properties for the material: $k = 0.12$, $\rho = 110 \text{ lb}_m$ per cu ft, and $c_p = 0.35$.

3. The slab of rubber of Ex. 13-2 is stripped from its mold at 350°F after the temperature is equalized. It is subsequently hung vertically in a room having air and surface temperatures of 80°F. Determine the temperature at the center of the slab after 12 min.

REFERENCES

1. Schneider, P. J.: "Conduction Heat Transfer," Addison-Wesley Publishing Company, Cambridge, Mass., 1955.
2. Carslaw, H. S.: "Introduction to the Mathematical Theory of Conduction of Heat in Solids," 2d ed., St. Martin's Press, Inc., New York, 1921.
3. Ingersoll, L. R., O. J. Zobel, and A. C. Ingersoll: "Heat Conduction with Engineering and Geological Applications," McGraw-Hill Book Company Inc., New York, 1948.
4. Williamson, E. D., and L. H. Adams: Temperature Distribution in Solids during Heating or Cooling, *Phys. Rev.*, **14**, 99–144 (1919).

CHAPTER 14

GRAPHICAL AND NUMERICAL METHODS
FOR HEAT-CONDUCTION PROBLEMS

14-1. General. In the design of the solid portions of various types of mechanical equipment, it is frequently necessary to determine maximum temperatures, heat-transfer rates, and forces or stresses produced by temperature. These effects are dependent upon temperature distribution. It is therefore necessary in such cases to describe in some manner the temperature field in terms of the space coordinates if steady state is involved or in terms of space and time coordinates if transient or cyclic conditions exist.

There are several methods available for describing temperature fields within a solid body. These are (1) analytical, (2) graphical, (3) numerical, (4) analog, and (5) experimental. The selection of one of these methods for a specific problem will depend on the existence of a satisfactory technique and the amount of time required to produce a satisfactory solution. Each method has advantages over the other methods for certain types of problems.

14-2. The Analytical Method. The determination of temperature fields by the analytical method involves a number of steps. First, a set of simplifying assumptions must be made regarding all the characteristics of the heat-transfer system, both within the system and at its boundaries. Next, it is necessary to derive a characteristic differential equation for the temperature as a function of the space coordinates or as a function of the space and time coordinates. Equation (13-1) is one form of such a differential equation for a specific set of simplifying assumptions. The differential equation must then be solved to give the temperature as a function of the coordinates. This solution must satisfy the boundary conditions or the boundary and the initial conditions if time is a coordinate. The resulting solution may then be used for computing the numerical values of the temperatures or heat-transfer rates.

The determination of the temperature field by this method cannot be considered "exact." In all cases the simplifying assumptions are such that the computed temperature distribution may be different from the actual distribution by small amounts or by large amounts, depending upon the validity of the assumptions. Nevertheless, this method is usually to be preferred in design problems since it lends itself most readily to a direct quantitative analysis of the effect of changing the various system characteristics such as dimensions and thermal quantities. Unfortunately, many real or practical boundary conditions are such that reasonably simple analytical solutions are not possible, although the use of new mathematical techniques and modern computing machines is making many more solutions possible than in the past.

14-3. Graphical Methods. Frequently because of the boundary conditions resulting from the geometry as well as the thermal situation at the boundaries, it is found that an analytical solution is not possible. In some such cases a solution which may be satisfactory for the purpose may be possible by a graphical method. In general, the graphical methods are somewhat more rapid but less precise than the analytical or numerical methods, even though the same simplifying assumptions regarding the system characteristics are made.

There are two basic graphical methods in common use for determining temperature fields or heat-transfer rates. These are the mapping method and the finite-difference-equation method. The mapping method which was first described by Lehmann[1] for electrical fields and later adapted by others to heat-conduction problems is useful for any two-dimensional steady-state heat-conduction problem where the boundary conditions can be readily determined but where the geometrical shape may make an analytical solution difficult and time-consuming.

This method is based upon the fact that a two-dimensional thermal field with constant thermal conductivity, and with no heat generation, is described by the Laplace equation. This equation is

$$\frac{\partial^2 \theta}{\partial x^2} + \frac{\partial^2 \theta}{\partial y^2} = 0 \qquad (14\text{-}1)$$

In this equation, which is a special form of Eq. (13-1), θ is the temperature at any point described by the space coordinates x and y. By using the theory of functions of complex variables, it can be shown that in a temperature field, described by Eq. (14-1), isothermal lines are everywhere at right angles to the adiabatic lines. Thus the heat flow which is in the direction of the adiabatic lines is in a direction perpendicular to the isothermal lines. These facts are utilized in the mapping process by sketching within a map of the two-dimensional space enclosed by the boundaries a network of curvilinear squares,* representing isothermal lines and channels of constant heat flow bounded by the adiabatic lines. The process usually requires that one starts from known isothermals and estimates the direction of equally spaced (temperaturewise) isothermals. Once these are established, curvilinear squares are formed by drawing lines at right angles to the isothermals forming figures that are approximately square.

To illustrate the use of this method a plane wall as described in Fig. 14-1b will be analyzed. The wall is assumed to be maintained at constant temperatures t_0 on one face and t_4 on the other face. A section of the wall S ft long and 1 ft deep will be analyzed. The thickness of the wall is L, and the thermal conductivity k is assumed constant.

The horizontal lines in the figure are isothermal lines. The temperature difference between two adjacent lines is Δt, i.e.,

$$\Delta t = (t_4 - t_3) = (t_3 - t_2) = (t_2 - t_1) = (t_1 - t_0) \qquad (14\text{-}2)$$

The vertical lines f_1, f_2, etc., are drawn perpendicular to the isothermal lines and form squares of width x and depth y. Since these are squares, $x = y$. The heat transfer across one square such as the one shown crosshatched is

$$q_i = k \frac{x}{y} \Delta t \qquad (14\text{-}3)$$

but since $x/y = 1$

$$q_i = k \Delta t \qquad (14\text{-}4)$$

---

* In a curvilinear square the adjacent sides form right angles and are approximately of equal mean length. As the number of such curvilinear squares in a g: en area is increased to infinity they approach true squares.

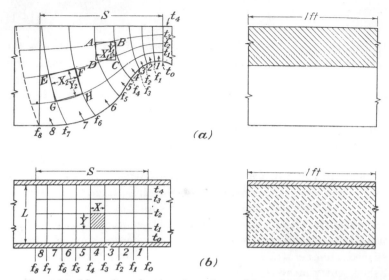

FIG. 14-1. Map of conduction through an object of irregular shape.

Since the channel formed by the lines f_3 and f_4 defines a channel of constant heat flow and since this is a steady-state problem, the rate of heat flow through channel 4 is

$$q_4 = k \, \Delta t \qquad (14\text{-}5)$$

By similar reasoning the rate of heat flow through the other full channels is $k \, \Delta t$; thus

$$q_1 = k \, \Delta t$$
$$q_2 = k \, \Delta t$$
$$\cdots \cdots \cdots \qquad (14\text{-}6)$$
$$q_7 = k \, \Delta t$$

The rate of heat flow through the partial channel 8 is

$$q_8 = rk \, \Delta t \qquad (14\text{-}7)$$

where r = fraction of a full channel represented by channel 8. The total rate of heat flow through the section of the wall being considered is

$$q = q_1 + q_2 + \cdots + q_8$$
or $\qquad q = Nk \, \Delta t + rk \, \Delta t$
hence $\qquad q = (N + r)k \, \Delta t \qquad (14\text{-}8)$

where N = number of full constant-heat-flow channels. To check the correctness of this we note that for this case the rate of heat flow can be expressed by the equation

$$q = \frac{kS(t_4 - t_0)}{L} \qquad (14\text{-}9)$$

But from the figure $S = (N + r)x$, $(t_4 - t_0) = n\,\Delta t$, and $L = ny$ where n = number of spaces between the isothermals. Hence from Eq. (14-9)

$$q = k\,\frac{(N + r)xn\,\Delta t}{ny}$$

or

$$q = (N + r)k\,\frac{x}{y}\,\Delta t$$

but $x/y = 1$. Thus

$$q = (N + r)k\,\Delta t$$

as given in Eq. (14-8).

Figure 14-1a is an illustration of the application of this method to an irregularly shaped wall. The lines t_0, t_1, t_2, t_3, and t_4 are isothermal lines with

$$\Delta t = (t_4 - t_3) = (t_3 - t_2) = (t_2 - t_1) = (t_1 - t_0)$$

The lines marked f_1, f_2, f_3, . . . , f_9 are lines of constant heat flow which form a network of curvilinear squares with the isothermal lines. The constant-heat-flow lines intersect the isothermals at right angles. It is to be noted that when the distance between isothermals is small, the distance between constant-heat-flow lines is likewise small, whereas when the distance between isothermals is large, the distance between constant-heat-flow lines is large.

The rate of heat transfer across any curvilinear square such as $ABCD$ may be written

$$q_{AB \to CD} = k\,\frac{x_1}{y_1}\,\Delta t$$

but in the curvilinear square $x_1 = y_1$, hence

$$q_{AB \to CD} = k\,\Delta t$$

Since the lines f_5 and f_4 form a constant-heat-flow channel and since the heat transfer is in steady state, the rate of heat transfer through the channel 5 is

$$q_5 = k \,\Delta t$$

By similar reasoning

$$q_{EF \to GH} = k \frac{x_2}{y_2} \,\Delta t$$

but since $x_2 = y_2$

$$q_{EF \to GH} = k \,\Delta t$$

and

$$q_7 = k \,\Delta t$$

Since for any of the curvilinear squares the dimension x is equal to the dimension y we write

$$q_1 = k \,\Delta t$$
$$q_2 = k \,\Delta t$$
$$\cdot\,\cdot\,\cdot\,\cdot\,\cdot\,\cdot\,\cdot$$
$$q_8 = r_1 k \,\Delta t$$

where r_1 = fraction of a complete channel formed by channel 8. Hence the total rate of heat transfer through the section of the wall being considered is

$$q = q_1 + q_2 + \cdots + q_8$$

or

$$q = (N + r_1)k \,\Delta t \qquad (14\text{-}8)$$

where N = number of full channels formed by the constant-heat-flow lines.

$\Delta t = (t_4 - t_0)/n$.

n = number of spaces between the isothermal lines.

Thus it is seen that this method of mapping may be used in the case of two-dimensional flow to determine the approximate heat flow between irregular surfaces. The accuracy of the method depends upon the size of the increment Δt. The smaller the increment the greater is the accuracy of mapping.

The graphical method for transient-heat-conduction problems is commonly known as the Schmidt-Binder[2,3] method, although several others have developed and improved it. This method is actually based on the replacement of the characteristic differential equation by a difference equation. The resulting equation may be used as the basis for some numerical-computation methods, and under certain circumstances it may be used as the basis for a graphical method. The use of the difference equation as the basis for the Schmidt-Binder graphical method for determining the temperature field in a one-dimensional transient-heat-conduction system is illustrated in the following.

Assume a one-dimensional heat-conduction system in which the temperature field is described by the following differential equation:

$$\alpha \frac{\partial^2 \theta}{\partial x^2} = \frac{\partial \theta}{\partial t} \qquad (14\text{-}10)$$

where θ = temperature.

α = thermal diffusivity.

x = distance in the direction of heat conduction.

t = time.

Let the time increments be Δt and the space increments be Δx. Then at a given point in a body a distance $m \, \Delta x$ from $x = 0$ and at a time $n \, \Delta t$ after time $t = 0$, the value of θ is $\theta_{m,n}$. At a time $(n + 1) \, \Delta t$, the temperature at this point would be $\theta_{m,n+1}$. Note that the subscript m refers to the space interval, while the subscript n refers to the time interval.

To replace the partial derivatives by finite difference ratios, it may be noted that

$$\frac{\partial \theta}{\partial x} \cong \frac{\theta_{m+1,n} - \theta_{m,n}}{\Delta x} \qquad \text{in the space interval } m \text{ to } m + 1$$

and $\quad \dfrac{\partial \theta}{\partial x} \cong \dfrac{\theta_{m,n} - \theta_{m-1,n}}{\Delta x} \qquad$ in the space interval $m - 1$ to m

Then $\qquad \dfrac{\partial^2 \theta}{\partial x^2} \cong \dfrac{1}{\Delta x} \left(\dfrac{\theta_{m+1,n} - \theta_{m,n}}{\Delta x} - \dfrac{\theta_{m,n} - \theta_{m-1,n}}{\Delta x} \right)$

or $\qquad \dfrac{\partial^2 \theta}{\partial x^2} \cong \dfrac{\theta_{m+1,n} - 2\theta_{m,n} + \theta_{m-1,n}}{(\Delta x)^2}$

also $\qquad \dfrac{\partial \theta}{\partial t} \cong \dfrac{\theta_{m,n+1} - \theta_{m,n}}{\Delta t}$

Hence Eq. (14-10) may be written

$$\alpha \frac{\theta_{m+1,n} - 2\theta_{m,n} + \theta_{m-1,n}}{(\Delta x)^2} = \frac{\theta_{m,n+1} - \theta_{m,n}}{\Delta t}$$

This equation may be solved for $\theta_{m,n+1}$, which is the future temperature excess at a time Δt after the present time $n \, \Delta t$. Thus

$$\theta_{m,n+1} = \frac{\theta_{m+1,n} + (M - 2)\theta_{m,n} + \theta_{m-1,n}}{M} \qquad (14\text{-}11)$$

where $\qquad M = \dfrac{(\Delta x)^2}{\alpha \, \Delta t} \qquad (14\text{-}12)$

If one starts with an initial temperature distribution, $\theta_{0,0}$, $\theta_{1,0}$, $\theta_{2,0}$, . . . , $\theta_{m,0}$, . . . , $\theta_{p,0}$, Eq. (14-11) may be used to compute the temperature distribution throughout the body at any time $n \, \Delta t$ by a step-by-step computation. The smallest possible effect $\theta_{m,n}$ could have on its own future value $\theta_{m,n+1}$ would be no effect. This would be the case if $M = 2$. For $M < 2$, the coefficient of θ would be negative. This would then mean that the larger $\theta_{m,n}$ is, the more it would reduce its own future temperature $\theta_{m,n+1}$. This is not consistent with the second law of thermodynamics. Furthermore, such a situation would result in oscillations of the value of θ at future times which might in some cases produce failure to converge.

If $M = 2$, then Eq. (14-11) becomes

$$\theta_{m,n+1} = \frac{\theta_{m+1,n} + \theta_{m-1,n}}{2} \qquad (14\text{-}13)$$

It may be seen by Eq. (14-13) that for $M = 2$, the future temperature at $m \, \Delta x$ is the average of the present temperatures at $(m - 1) \, \Delta x$ and $(m + 1) \, \Delta x$. This equation is the basis for the Binder-Schmidt graphical method for determining the temperature as a function of x for any successive instants of time differing by finite increments Δt. This method is illustrated in Fig. 14-2 for an infinite slab of finite thickness having an initial temperature distribution θ_{ia}, θ_{ib} with the surface temperatures suddenly changed to θ_0. To determine the new temperature at any time such as $\theta_{3,5}$ it is simply necessary to draw a line joining $\theta_{2,4}$ to $\theta_{4,4}$. The ordinate for the intersection of this line with the line $x = 3 \, \Delta x$ is the new temperature $\theta_{3,5}$. Of course, to arrive at the values of $\theta_{2,4}$ and $\theta_{4,4}$, one must start with the initial temperatures and determine the temperature distribution for the successive time intervals $n = 1$, $n = 2$, $n = 3$, and $n = 4$.

14-4. Numerical Methods (General). In the analytical method a set of characteristic differential equations is developed for each physical subdivision in a heat-transfer system and the temperature and its gradients at *any* point can be calculated, provided a solution can be found that satisfies the boundary and initial conditions. In the numerical method, however, it is assumed that the system can be subdivided into suitable regions for which a single temperature at some point in space or space and time is said to be representative of that region. The

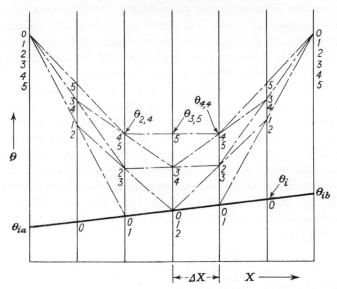

FIG. 14-2. Schmidt-Binder graphical method applied to one-dimensional hea.
conduction for case of infinitely wide slab of finite thickness with constant
conductivity and no heat generation. Initial temperature distribution
θ_i with surface temperatures suddenly changed to and maintained at θ_0.
Note that the numbers indicate the time interval for which the temperature
applies. Each time interval Δt is equal to $(\Delta x)^2/2\alpha$, and the total elapsed
time is $n\,\Delta t$, where n is the number indicated in the sketch.

temperatures are then computed for these preselected, discrete
points. This is quite analogous to replacing the differential
equation by a difference equation. In many cases an exactly
equivalent equation results for both methods. Since it is axio-
matic that the larger the number of subdivisions used, the greater
will be the precision but also the greater will be the number of
computations required, it therefore follows that the number of
subdivisions used must be as small as possible, consistent with
the desired precision.

This method may be applied either to a homogeneous region
in a heat-transfer system, or it may be applied to a set of phys-
ically different regions that are connected thermally.*   In either
case the method used to develop the general equations will be
the same. The actual final computation method selected will be

---

* Thermally connected regions are regions which can interchange heat
with each other by virtue of a temperature difference existing between them.

different for different types of problems, the objective being to select the method which will result in the smallest number of computations to arrive at the desired result.   The general equation for use in numerical computations will be developed here. Its application to some special cases is considered in subsequent articles.   For a more extensive treatment of this subject the reader may refer to Dusinberre's[4] excellent text.

Assume a heat-transfer system subdivided into n regions which are thermally connected.   A heat balance for each subdivision may be written.   If we assume that a time interval Δt sufficiently small is selected so that the temperature differences at the beginning of the time interval and the temperature differences at the end of the time interval are not sufficiently different to produce a significant error, then each heat-balance equation will have the form

$$Q_{g0} + \sum_{i=1}^{i=n} K_{i0}(\theta_i - \theta_0) = \frac{C_0}{\Delta t}(\theta_0' - \theta_0) \qquad (14\text{-}14)$$

where the subscript 0 refers to the subdivision for which the heat-balance equation is written and the subscript i refers to the subdivisions which are thermally connected to it.   The symbols in Eq. (14-14) have the following meanings:

Q_{g0} = heat generated per unit time within the subdivision 0.

K_{i0} = thermal conductance between subdivision i and subdivision 0.   This is the time rate of heat transfer per unit temperature difference.

θ_i = representative temperature of subdivision i at the beginning of the time interval.

θ_0 = representative temperature of the subdivision 0 at the beginning of the time interval.

θ_0' = representative temperature of subdivision 0 at the end of the time interval.

C_0 = thermal capacitance of subdivision 0.   This is the heat-storage capacity per unit of temperature change.

Δt = time interval.

Equation (14-14) may be solved for θ_0'.   Thus

$$\theta_0' = \frac{Q_{g0}\,\Delta t}{C_0} + \frac{\Delta t}{C_0}\sum_{i=1}^{i=n} K_{i0}\theta_i + \theta_0\left(1 - \frac{\Delta t}{C_0}\sum_{i=1}^{i=n} K_{i0}\right) \quad (14\text{-}15)$$

For each subdivision a heat-balance equation such as Eq. (14-14) may be written. The process of calculation then consists of computing the new temperatures θ_0' for each region from the temperature distribution at the beginning of the time interval. The new set of temperatures will then be the temperatures to be used for the next time interval. Equation (14-14) may be used for either steady-state or transient computations. A different calculation procedure will be required for the two cases.

In the steady-state case $\theta_0' = \theta_0$, therefore, the right-hand side of Eq. (14-14) is zero and the problem is to determine the temperatures at each subdivision from the boundary conditions. For the transient case the representative temperatures for each subdivision at the time $t = 0$ must be known or they must be determined before subsequent temperatures may be calculated.

14-5. Numerical Methods (Steady-state Conduction). To illustrate the use of Eq. (14-14) for a heat-transfer system in which the process is one of conduction heat transfer within the boundaries of a homogeneous body, let us assume that the body can be divided into a network of geometrical subdivisions and that the representative temperature of each subdivision is the temperature at its center. For example, the network chosen may be square for a case of two-dimensional steady-state heat conduction as illustrated in Fig. 14-3. It will be noticed that the subdivision 0 is thermally connected only with the subdivisions 1, 2, 3, and 4. Subdivisions 5, 6, 7, and 8 are not thermally connected with

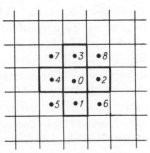

Fig. 14-3. Square network applied to a corner.

subdivision 0 since there is no area in common through which heat may be transferred. Thus the heat balance of Eq. (14-14) when applied to subdivision 0 becomes

$$Q_{q0} + K_{10}\theta_1 + K_{20}\theta_2 + K_{30}\theta_3 + K_{40}\theta_4$$
$$- \theta_0(K_{10} + K_{20} + K_{30} + K_{40}) = 0$$

In this case if we assume that the thermal conductivity k is constant, we may replace the K_{i0} terms by

$$\frac{k\,\overline{\Delta x}\,b}{\Delta x} = kb$$

since the area through which heat is conducted is $\Delta x\, b$, where b is the thickness perpendicular to the plane of the paper, and the length of the path for heat conduction is $\overline{\Delta x}$. Thus the heat-balance equation may be written

$$Q_{g0} + kb(\theta_1 + \theta_2 + \theta_3 + \theta_4 - 4\theta_0) = 0$$

or $$\theta_1 + \theta_2 + \theta_3 + \theta_4 - 4\theta_0 = \frac{Q_{g0}}{kb} \qquad (14\text{-}16)$$

If it is assumed that Q_{g0}, the heat generated, is zero, the computation equation becomes

$$\theta_1 + \theta_2 + \theta_3 + \theta_4 - 4\theta_0 = 0 \qquad (14\text{-}17)$$

This equation may be used as the basis for a relaxation method of computation. This method would consist of evaluating an equation similar to Eq. (14-17) for each subdivision in the network including those at the boundaries. Any convenient set of representative temperatures for each subdivision may be assumed at the start. With this set of representative temperatures, the left-hand side of Eq. (14-17) is evaluated for each subdivision. In general this will not be equal to zero for all subdivisions. Thus a set of residuals R_0 [the amount by which the left-hand side of Eq. (14-17) is different from zero] will be computed for each subdivision. The temperature θ_0, of the subdivision for which the residual R_0 is different from zero by the largest amount, must then be corrected by an amount $R_0/4$. The other temperatures are not changed for the next computation of residuals. When the values of R_0 are computed for the changed temperature, a new correction is made. This process is repeated until all residuals are as close to zero as is desired.

To illustrate this method the two-dimensional heat-transfer system in Fig. 14-4 is analyzed for the case where the boundary temperatures are known and there is no heat generation.* It will be noted that a square network is assumed with the representative temperature for each subdivision taken as the temperature at its center. It will also be noted that at the boundaries half squares are assumed so that the boundary temperature will be the representative temperature for those subdivisions.

---

* This problem was originally described by H. W. Emmons in *Trans. ASME*, **65**, 607–612 (1943).

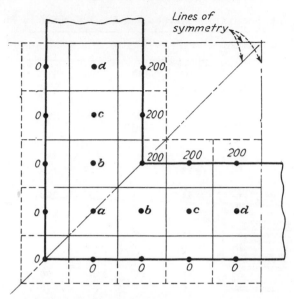

Fig. 14-4. Square network for corner of infinite thickness in direction perpendicular to plane of paper (no heat transfer in direction perpendicular to paper). Inside-surface temperature = 200°F. Outside-surface temperature = 0°F. Square network used to determine θ_a, θ_b, θ_c, θ_d. No heat generation. [*H. W. Emmons, Trans. ASME*, **65**, 607–612 (1943).]

Table 14-1 shows the relaxation method of computation used for the determination of the temperatures at a, b, c, and d.

Sample computation for line (1):

1. Assume $\theta_a = \theta_b = \theta_c = \theta_d = 100$.
2. Heat balances [Eq. (14-17)].

$$a \begin{cases} \theta_b + \theta_b + 0 + 0 - 4\theta_a = R_a \\ 100 + 100 - 4 \times 100 = R_a = -200 \end{cases}$$

$$b \begin{cases} \theta_a + 200 + \theta_c + 0 - 4\theta_b = R_b \\ 100 + 200 + 100 - 4 \times 100 = R_b = 0 \end{cases}$$

$$c \begin{cases} \theta_b + 200 + \theta_d + 0 - 4\theta_c = R_c \\ 100 + 200 + 100 - 400 = R_c = 0 \end{cases}$$

$$d \begin{cases} \theta_c + 200 + \theta_c - 4\theta_d = R_d \\ 100 + 200 + 100 - 400 = R_d = 0 \end{cases}$$

3. Correction. Since R_a is the largest residual, then the value of θ_a should be corrected by the amount $R_a/4$. Thus the new θ_a will be $100 - (^{200}\!/_4) = 50$.

TABLE 14-1. RELAXATION METHOD FOR TEMPERATURES IN FIG. 14-4

Position		a	b	c	d
1	θ	100	100	100	100
	R	-200	0	0	0
2	θ	50	100	100	100
	R	0	-50	0	0
3	θ	50	88	100	100
	R	-24	-2	-12	0
4	θ	44	88	100	100
	R	0	-8	-12	0
5	θ	44	88	97	100
	R	0	-11	0	-6
6	θ	44	85	97	100
	R	-6	$+1$	-3	-6
7	θ	42	85	97	100
	R	$+2$	-1	-3	-6
8	θ	42	85	97	98
	R	$+2$	-1	-5	$+2$
9	θ	42	85	96	98
	R	$+2$	-2	-1	0

4. New temperatures for next computation: $\theta_a = 50$, $\theta_b = 100$, $\theta_c = 100$, $\theta_d = 100$.

5. Repeat calculations until corrections are within the desired limits. In this case line 9 shows that no residual is greater than 2, therefore no correction would be greater than 0.5°. Hence the temperatures in line 9 are within 0.5° of the correct values. The temperatures in line 9 could be used as the basis for starting a finer network if temperatures at additional points are desired.

If the boundary temperatures were not known but the temperatures of the fluids in contact with the boundaries were given and the heat-transfer coefficients at the boundaries could be computed, then heat balances for the subdivisions at the boundaries would need to be included, with the Q_{g0}/kb term in Eq. (14-16) replaced by a term including an unknown temperature and the heat-transfer coefficient.   Thus for any of the subdivisions at the boundaries, immediately below the subdivision marked a in Fig. 14-4,

$$\frac{Q_{g0}}{kb} = \frac{h\,\Delta x\,b(\theta_a - \theta_0)}{kb} = \frac{h\,\Delta x}{k}\,(\theta_e - \theta_0)$$

where θ_e = temperature of environment.   Hence Eq. (14-16) would read

$$\theta_1 + \theta_2 + \theta_3 + \frac{h\,\Delta x}{k}\,\theta_e - \left(3 + \frac{h\,\Delta x}{k}\right)\theta_c = 0 \qquad (14\text{-}18)$$

The method used in the case of the corner pictured in Fig. 14-4 can be extended to the three-dimensional case by taking into account the fact that each region in a cubical network would be surrounded by six regions thermally connected with it.   Other types of networks may be more convenient for shapes with boundaries that are not at right angles or parallel to each other.

14-6. Numerical Methods (Transient Conduction).   For transient problems it is necessary to apply Eq. (14-15) to each subdivision of the region.   Any convenient network of subdivisions may be used.   It will be recalled that there are two types of subdivision for this calculation.   These are the space subdivision and the time subdivision.   Thus a calculation procedure different from the relaxation method is required.

The heat-balance equation (14-15) may be written for the two-dimensional space shown in Fig. 14-3 with the assumptions that the thermal conductivity k is constant, the specific heat is c, and the density is ρ.   With these assumptions the following substitutions are made:

$$Q_{g0} = 0$$

$$K_{i0} = \frac{kb\,\Delta x}{\Delta x} = kb$$

$$C_0 = \rho cb\,\overline{\Delta x^2}$$

$$\alpha = \frac{k}{c\rho}$$

and Eq. (14-15) becomes

$$\theta_0' = \frac{\alpha\,\Delta t}{\Delta x^2}\sum_{i=1}^{i=n}\theta_i + \theta_0\left(1 - \frac{\alpha\,\Delta t}{\Delta x^2}\,n\right) \qquad (14\text{-}19)$$

where n is the number of subdivisions in thermal contact with region 0. In the two-dimensional case of Fig. 14-3, n would be 4. For a one-dimensional case n is equal to 2. Examination of Eq. (14-19) is possible for the selection of values for the size of the time increment Δt and the space increment Δx. The criteria are that Δt and Δx should be as large as possible to keep the number of computations at a minimum consistent with the desired precision, but at the same time the values selected must not be such as to go contrary to the second law of thermodynamics nor such as to produce oscillations of the value of θ_0'. The last term in Eq. (14-19) shows that the smallest value possible for $[1 - n(\alpha \, \Delta t)/\overline{\Delta x}^2]$ is zero. Even this value would say that the present temperature θ_0 has no effect on its own future temperature θ_0'. In general, therefore,

$$\frac{\alpha \, \Delta t}{\overline{\Delta x}^2}n \le 1$$

or
$$\Delta t \le \frac{\overline{\Delta x}^2}{n\alpha} \tag{14-20}$$

If the term $(\alpha \, \Delta t)/\overline{\Delta x}^2$ is replaced by $1/M$, Eq. (14-19) becomes

$$\theta_0' = \frac{1}{M} \left[\sum_{i=1}^{i=n} \theta_i + \theta_0(M - n) \right] \tag{14-21}$$

and Eq. (14-20) tells us that

$$M \ge n \tag{14-22}$$

For one-dimensional heat transfer $(n = 2)$, it is apparent that Eq. (14-21) is the same as Eq. (14-11) which was developed from the difference equations.

To illustrate a calculation procedure which may be used for transient-heat conduction, the case of an infinitely thick plate initially having a surface temperature $\theta_0 = 0$ with the surface temperature instantaneously changed to $\theta = 100$ F is considered. This is a one-dimensional heat-conduction problem, and the region may be subdivided as shown in Fig. 14-5. (Note that a subdivision of width $\Delta x/2$ is taken adjacent to the surface so that the surface temperature is at the center of a fictitious region of width Δx.)

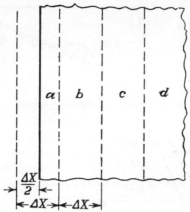

FIG. 14-5. Subdivisions for one-dimensional system.

Equation (14-21) for the case where $M = n = 2$ will be

$$\theta_0' = \frac{1}{2} \sum_{i=1}^{i=2} \theta_i$$

Hence for the new temperature at a sample point b, the equation is

$$\theta_b' = \frac{\theta_a + \theta_c}{2}$$

Table 14-2 shows the procedure for calculating the temperatures at various times. In this calculation it is assumed that

$$M = 2 \qquad \frac{\overline{\Delta x}^2}{\alpha} = 1 \qquad \Delta t = \tfrac{1}{2} \text{ hr}$$

TABLE 14-2

Time, hr	Subdivision temperatures, °F				
	a	b	c	d	e
0.0	100	0	0	0	0
0.5	100	50	0	0	0
1.0	100	50	25	0	0
1.5	100	62.5	25	12.5	0
2.0	100	62.5	37.5	12.5	0
2.5	100	68.8	37.5	18.8	6.3
3.0	100	68.8	43.8	21.9	9.4

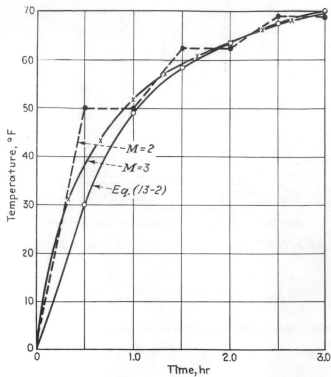

FIG. 14-6. Time-temperature curves from Tables 14-2 and 14-3.

Table 14-3 shows the values computed for the same region but with $M = 3$; $\overline{\Delta x}^2/\alpha = 1$; $\Delta t = \frac{1}{3}$ hr. It will be noted that for this case Eq. (14-21) becomes

$$\theta_0' = \frac{1}{M} \sum_{i=1}^{i=2} \theta_i + \frac{1}{M} \theta_0$$

and it can be seen that the new temperature for the sample point b is

$$\theta_b' = \frac{\theta_a + \theta_b + \theta_c}{3}$$

The temperature for point b is plotted in Fig. 14-6 for the values from Tables 14-2 and 14-3 and as computed from Eq. (13-2) for purposes of comparison. It will be noted that the three curves

TABLE 14-3

Time, hr	Subdivision temperatures, °F				
	a	b	c	d	e
0.0	100	0	0	0	0
0.33	100	33.3	0	0	0
0.67	100	44.4	11.1	0	0
1.00	100	51.9	18.5	3.7	0
1.33	100	56.8	24.7	7.4	1.2
1.67	100	60.5	29.6	11.1	2.9
2.00	100	63.4	33.7	14.5	4.8
2.33	100	65.7	37.2	18.0	7.2
2.67	100	67.7	40.3	20.8	9.2
3.00	100	69.3	43.9	22.4	11.1

approach each other after very long times. However, it is interesting that the value $M = 3$, after a time between 1.0 and 1.5 hr, gives results which are very close to the analytical result based on Eq. (13-2). Computations for a value of $M = 4$ would approach the analytical result even more closely.

Problems

1. Find, by mapping, the heat flow from the upper to the lower surface of the object shown below. Assume that no heat is lost from the sides. k for the material equals 11 Btu/(hr)(sq ft)(°F/ft).

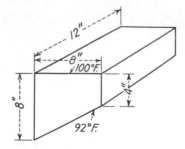

2. A flat wall 6 in. thick is initially at a uniform temperature of 0°F The temperatures of the two surfaces are suddenly changed to 100°F and maintained at that value. Use the graphical method described in Art. 14-3 to find the temperature at the end of 5 min for 1-in. intervals. Assume thermal diffusivity α is $\frac{5}{24}$ (sq ft)/(hr).

3. A long thick-walled square tube has an inside edge that is 12 in. and an outside edge that is 36 in. The inside-surface temperature is 500°F,

and the outside is 100°F. Using the relaxation method described in Art. 14-5 determine the temperatures for points 6 in. apart halfway between the inside and outside.

4. Solve Prob. 2 using the iterative method described in Art. 14-6. Use $M = 4$. Compare the result with the result of Prob. 2 and with the result obtained by using Eq. (13-6).

REFERENCES

1. Lehmann, T.: *Elektrotech. Z.*, **30**, 995, 1019 (1909).
2. Binder, L.: Dissertation, Technisch Hochschule, Muenchen, W. Knapp, Halle, 1911.
3. Schmidt, E., "Foppls Festschrift," pp. 179–198, Springer-Verlag OHG, Vienna, 1924.
4. Dusinberre, G. M., "Numerical Analysis of Heat Flow," McGraw-Hill Book Company, Inc., New York, 1949.

APPENDIX

DERIVATION OF STEFAN-BOLTZMANN LAW FROM PLANCK'S EQUATION

Planck's equation for the distribution of radiant energy is, according to Eq. (4-7),

$$E_\lambda = \frac{1.16 \times 10^8 \lambda^{-5}}{e^{25,740/\lambda T} - 1}$$

The total energy of radiation for a black body over the entire range of wavelengths from $\lambda = 0$ to $\lambda = \infty$ is

$$E = \int_0^\infty \frac{1.16 \times 10^8 \lambda^{-5}}{e^{25,740/\lambda T} - 1} \, d\lambda$$

If $A = 1.16 \times 10^8$, $B = 25,740/T$, and $y = B/\lambda$, then $\lambda = B/y$ and $d\lambda = -(B/y^2) \, dy$; hence,

$$E = -\int_\infty^0 \frac{A}{B^4} y^3 (e^y - 1)^{-1} \, dy = \frac{A}{B^4} \int_0^\infty y^3 (e^y - 1)^{-1} \, dy$$

By expanding the term in the parentheses, this becomes

$$E = \frac{A}{B^4} \int_0^\infty y^3 (e^{-y} + e^{-2y} + e^{-3y} + e^{-4y} + \cdots) \, dy$$

which can be integrated term by term, since any complete table of integrals will show that $\int_0^\infty y^n e^{-ay} \, dy = n!/a^{n+1}$. Therefore

$$E = \frac{A}{B^4} \left(\frac{3!}{1^4} + \frac{3!}{2^4} + \frac{3!}{3^4} + \frac{3!}{4^4} + \cdots \right)$$

$$= \frac{A}{B^4} \times 6 \left(\frac{1}{1^4} + \frac{1}{2^4} + \frac{1}{3^4} + \frac{1}{4^4} + \cdots \right)$$

Neglecting all other terms beyond $1/4^4$ and replacing A and B by their proper values, this becomes

$$E = \frac{6.45 \times 1.16 \times 10^8 \times T^4}{25,740^4}$$

$$= 0.173 \times 10^{-8} \times T^4$$

DERIVATION OF EQUATION OF NET HEAT-RADIATION EXCHANGE BETWEEN TWO SURFACES THAT ARE "NONBLACK"

Assume two surfaces of area A_1 and A_2 at temperatures T_1 and T_2.

Let ϵ_1 and ϵ_2 = emissivity of surface A_1 and A_2, respectively.

E_1 and E_2 = total emissive power of A_1 and A_2, respectively.

r_1 and r_2 = reflectivity of A_1 and A_2, respectively

$$(r_1 = 1 - \epsilon_1 \text{ and } r_2 = 1 - \epsilon_2).$$

f_2 = fraction of energy leaving A_1 and falling on A_2.

f_1 = fraction of energy leaving A_2 and falling on A_1.

The energy that originates at A_1 and falls on A_2 then is $\epsilon_1 E_1 A_1 f_2$. The surface A_2 absorbs only a portion of this and reflects the amount $r_2\epsilon_1 E_1 A_1 f_2$, and the fraction of this reflected energy which falls on A_1 is $r_2\epsilon_1 E_1 A_1 f_2 f_1$. The surface A_1 in turn will reflect a portion of this equal to $r_1 r_2\epsilon_1 E_1 A_1 f_2 f_1$, and the fraction of this energy which falls on A_2 is $r_1 r_2\epsilon_1 E_1 A_1 f_2^2 f_1$. The process goes on indefinitely, and it can be seen that the energy that originates at the surface A_1 and falls on the surface A_2 both directly and by repeated reflection of the original energy, $\epsilon_1 E_1 A_1 f_2$, is

$$q_a = \epsilon_1 E_1 A_1 f_2 + r_1 r_2 f_1 f_2^2 \epsilon_1 E_1 A_1 + r_1^2 r_2^2 f_1^2 f_2^3 \epsilon_1 E_1 A_1 + \cdots \quad (A\text{-}1)$$

However, this is only a part of the total energy that falls on the surface A_2, for it must be remembered that part of the energy originating in the body 2 will be returned to it by reflection from surface A_1. Thus the energy that originates at A_2 and falls on A_1 is $\epsilon_2 E_2 A_2 f_1$. The portion of this which is reflected by A_1 and again strikes A_2 is $r_1\epsilon_2 E_2 A_2 f_1 f_2$, whereupon the amount $r_1 r_2\epsilon_2 E_2 A_2 f_1^2 f_2$ is returned to A_1. The fraction of this energy which is reflected by A_1 and again falls on A_2 is $r_1^2 r_2\epsilon_2 E_2 A_2 f_1^2 f_2^2$. This process also goes on indefinitely. Therefore the portion of the energy that originates at A_2 and is returned to it by repeated reflections is

$$q_b = r_1\epsilon_2 E_2 A_2 f_1 f_2 + r_1^2 r_2\epsilon_2 E_2 A_2 f_1^2 f_2^2 + r_1^3 r_2^2 \epsilon_2 E_2 A_2 f_1^3 f_2^3 + \cdots \quad (A\text{-}2)$$

Therefore the total energy that leaves the surface A_1 and falls on the surface A_2 is $q_{1-2} = q_a + q_b$, or

$$q_{1-2} = \epsilon_1 E_1 A_1 f_2 (1 + r_1 r_2 f_1 f_2 + r_1^2 r_2^2 f_1^2 f_2^2 + \cdots)$$
$$+ \epsilon_2 E_2 A_2 r_1 f_1 f_2 (1 + r_1 r_2 f_1 f_2 + r_1^2 r_2^2 f_1^2 f_2^2$$
$$+ r_1^3 r_2^3 f_1^3 f_2^3 + \cdots) \quad \text{(A-3)}$$

This may be written

$$q_{1-2} = (\epsilon_1 E_1 A_1 f_2 + \epsilon_2 E_2 A_2 r_1 f_1 f_2)(1 + r_1 r_2 f_1 f_2 + r_1^2 r_2^2 f_1^2 f_2^2$$
$$+ r_1^3 r_2^3 f_1^3 f_2^3 + \cdots) \quad \text{(A-4)}$$

It can be seen that the term in the second parentheses is a geometric series having the common multiplier $r_1 r_2 f_1 f_2$. This may be summed up to yield the expression

$$q_{1-2} = \frac{\epsilon_1 E_1 A_1 f_2 + \epsilon_2 E_2 A_2 r_1 f_1 f_2}{1 - r_1 r_2 f_1 f_2} \quad \text{(A-5)}$$

By a similar process of reasoning the expression for the total energy that leaves the surface A_2 and falls on A_1 is found to be

$$q_{2-1} = \frac{\epsilon_2 E_2 A_2 f_1 + \epsilon_1 E_1 A_1 r_2 f_1 f_2}{1 - r_1 r_2 f_1 f_2} \quad \text{(A-6)}$$

The net heat-radiation exchange between the two surfaces is, therefore,

$$q = q_{1-2} - q_{2-1}$$

or

$$q = \frac{\epsilon_1 E_1 A_1 f_2 - \epsilon_2 E_2 A_2 f_1 + \epsilon_2 E_2 A_2 r_1 f_1 f_2 - \epsilon_1 E_1 A_1 r_2 f_1 f_2}{1 - r_1 r_2 f_1 f_2} \quad \text{(A-7)}$$

If E_1 is replaced by σT_1^4 and E_2 is replaced by σT_2^4, then

$$q = \frac{\sigma A_1 \left(\epsilon_1 T_1^4 f_2 - \epsilon_2 T_2^4 f_1 \dfrac{A_2}{A_1} + r_1 \epsilon_2 T_2^4 \dfrac{A_2}{A_1} f_1 f_2 - r_2 \epsilon_1 T_1^4 f_1 f_2 \right)}{1 - r_1 r_2 f_1 f_2}$$

which is the same as Eq. (4-26).

DERIVATION OF EQUATION OF NET RADIATION HEAT EXCHANGE BETWEEN TWO SURFACE ELEMENTS THAT ARE "BLACK BODIES"

If we visualize the radiation coming from the element dA_1 of Fig. 4-5, page 59, as lines emanating from dA_1, then the lines will be closer together in the direction of ON than in any other direction. Hence, there would be more lines per unit surface area of the hemisphere at N than at any other position such as at P. This is known as Lambert's cosine law.

If we call i the intensity of radiation falling on any elemental area on the surface of the hemisphere such as at dS, then the total energy falling on the surface of the hemisphere will be

$$dq_1 = \int_s i \, dS \qquad \text{(A-8)}$$

where the integral is taken over the entire surface of the hemisphere. But by Lambert's cosine law

$$i = i_N \cos \theta_1 \qquad \text{(A-9)}$$

where i_N is the intensity of radiation in the direction of the normal. Furthermore, from the accompanying figure we see that

$$ds = r \sin \theta_1 \, d\phi \, r \, d\theta_1 \qquad \text{(A-10)}$$

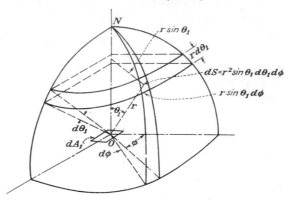

Hence when the values of i and ds from Eqs. (A-9) and (A-10) are substituted into Eq. (A-8) we get

$$dq_1 = \int_{\phi=0}^{\phi=2\pi} \int_{\theta_1=0}^{\theta_1=\pi/2} i_N \cos \theta_1 \sin \theta_1 r^2 \, d\theta_1 \, d\phi$$

or
$$dq_1 = \pi r^2 i_N \qquad \text{(A-11)}$$

This is the total radiant energy leaving the surface dA_1. Equation (A-11) may be written

$$i_N = \frac{dq_1}{\pi r^2}$$

Substituting this into Eq. (A-9) we get

$$i = \frac{dq_1}{\pi r^2} \cos \theta_1$$

which gives the energy falling per unit area of the hemisphere at any point P. The energy falling on the element dS will be $dq_1 (\cos \theta_1/\pi r^2) \, dS$, but for dS we may substitute $\cos \theta_2 \, dA_2$; hence the portion of the total energy leaving dA_1 and falling on dA_2 will be

$$dq_{1-2} = dq_1 \frac{\cos \theta_1 \cos \theta_2 \, dA_2}{\pi r^2}$$

Thus the fraction of the energy leaving dA_1 and falling on dA_2 is

$$f_2 = \frac{\cos \theta_1 \cos \theta_2 \, dA_2}{\pi r^2}$$

By exactly similar reasoning

$$f_1 = \frac{\cos \theta_1 \cos \theta_2 \, dA_1}{\pi r^2}$$

Equations (4-20) and (4-21) on page 60 may now be written

$$dq_{1-2} = \frac{\sigma \cos \theta_1 \cos \theta_2 \, dA_2 \, dA_1 \, T_1{}^4}{\pi r^2}$$

$$dq_{2-1} = \frac{\sigma \cos \theta_1 \cos \theta_2 \, dA_2 \, dA_1 \, T_2{}^4}{\pi r^2}$$

or the net heat exchange between the two surface elements is

$$dq = dq_{1-2} - dq_{2-1} = \frac{\sigma \cos \theta_1 \cos \theta_2}{\pi r^2} dA_2 \, dA_1 \, (T_1{}^4 - T_2{}^4)$$

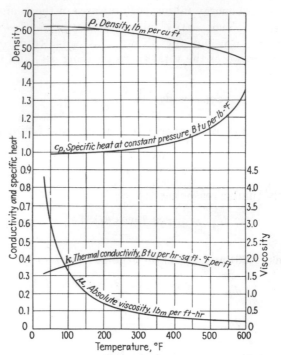

FIG. A-1. Properties of water.

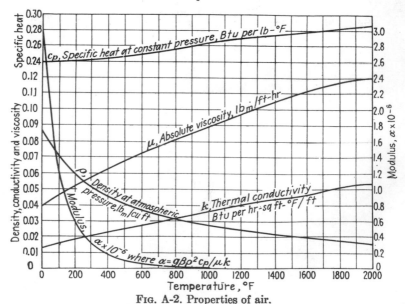

FIG. A-2. Properties of air.

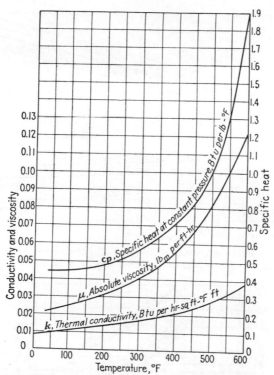

FIG. A-3. Properties of saturated steam. [*Values of thermal conductivity from F. G. Keyes and D. J. Sandell, Jr., New Measurements of the Heat Conductivity of Steam and Nitrogen, Trans. ASME,* **72,** 767 (1950).]

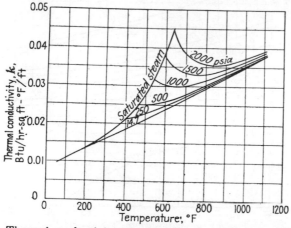

FIG. A-4. Thermal conductivity of superheated steam. [*Values of thermal conductivity from F. G. Keyes and D. J. Sandell, Jr., New Measurements of the Heat Conductivity of Steam and Nitrogen, Trans. ASME,* **72,** 767 (1950).]

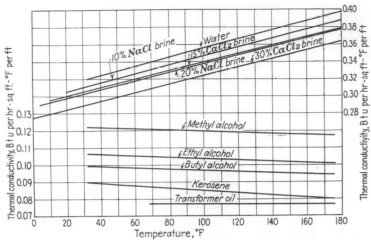

FIG. A-5. Thermal conductivity of liquids. ("*International Critical Tables*," *McGraw-Hill Book Company, Inc., New York, 1926–1930.*)

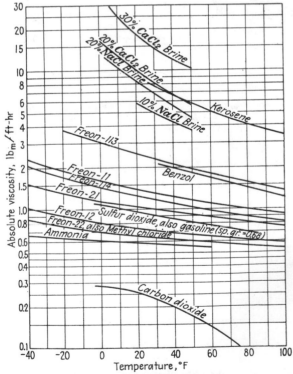

FIG. A-6. Viscosities of liquids.

TABLE A-1. PROPERTIES OF WATER

°F	c_p, $\dfrac{\text{Btu}}{(\text{lb}_m)(°\text{F})}$	ρ, $\dfrac{\text{lb}_m}{\text{cu ft}}$	μ, $\dfrac{\text{lb}_m}{(\text{ft})(\text{hr})}$	k, $\dfrac{\text{Btu}}{(\text{hr})(\text{ft})(°\text{F})}$	$\dfrac{c_p\mu}{k}$	a, $\dfrac{1}{(\text{cu ft})(°\text{F})}$
32	1.009	62.42	4.33	0.327	13.35	
40	1.005	62.42	3.75	0.332	11.35	0.3×10^8
50	1.002	62.38	3.17	0.338	9.40	1.0×10^8
60	1.000	62.34	2.71	0.344	7.88	1.7×10^8
70	0.998	62.27	2.37	0.349	6.78	2.3×10^8
80	0.998	62.17	2.08	0.355	5.85	3.0×10^8
90	0.997	62.11	1.85	0.360	5.12	3.9×10^8
100	0.997	61.99	1.65	0.364	4.53	5.2×10^8
110	0.997	61.84	1.49	0.368	4.04	6.6×10^8
120	0.997	61.73	1.36	0.372	3.64	7.7×10^8
130	0.998	61.54	1.24	0.375	3.30	8.9×10^8
140	0.998	61.39	1.14	0.378	3.01	10.2×10^8
150	0.999	61.20	1.04	0.381	2.73	12.0×10^8
160	1.000	61.01	0.97	0.384	2.53	13.9×10^8
170	1.001	60.79	0.90	0.386	2.33	15.5×10^8
180	1.002	60.57	0.84	0.389	2.16	17.1×10^8
190	1.003	60.35	0.79	0.390	2.03	
200	1.004	60.13	0.74	0.392	1.90	
220	1.007	59.63	0.65	0.395	1.66	
240	1.010	59.10	0.59	0.396	1.51	
260	1.015	58.51	0.53	0.396	1.36	
280	1.020	57.94	0.48	0.396	1.24	
300	1.026	57.31	0.45	0.395	1.17	
350	1.044	55.59	0.38	0.391	1.02	
400	1.067	53.65	0.33	0.384	1.00	
450	1.095	51.55	0.29	0.373	0.85	
500	1.130	49.02	0.26	0.356	0.83	
550	1.200	45.92	0.23			
600	1.362	42.37	0.21			

Sources of data:

c_p—ASHVE Guide, 1940.

ρ—Keenan, J. H., and F. G. Keyes, "Thermodynamic Properties of Steam," John Wiley & Sons, Inc., New York, 1936.

μ—International Critical Tables.

k—Schmidt, E., and W. Selschopp, "Forschung auf dem Gebiete des Ingenieurwesens," **3**, 277, 1932.

Notation: $a = g\beta\rho^2 c_p/\mu k$.

Table A-2. Properties of Air

°F	c_p, Btu $\overline{(\mathrm{lb}_m)(°F)}$	ρ,* lb_m $\overline{\mathrm{cu\ ft}}$	μ, lb_m $\overline{(\mathrm{ft})(\mathrm{hr})}$	k, Btu $\overline{(\mathrm{hr})(\mathrm{ft})(°F)}$	$\dfrac{c_p\mu}{k}$	a,* 1 $\overline{(\mathrm{cu\ ft})(°F)}$
0	0.239	0.0862	0.040	0.0132	0.72	3.0×10^6
20	0.240	0.0826	0.041	0.0138	0.71	2.5×10^6
40	0.240	0.0793	0.043	0.0143	0.71	2.1×10^6
60	0.240	0.0763	0.044	0.0148	0.71	1.7×10^6
80	0.240	0.0734	0.045	0.0153	0.70	1.4×10^6
100	0.240	0.0708	0.046	0.0158	0.70	1.2×10^6
120	0.240	0.0684	0.047	0.0162	0.70	1.1×10^6
140	0.240	0.0661	0.049	0.0168	0.70	0.89×10^6
160	0.241	0.0640	0.050	0.0172	0.70	0.77×10^6
180	0.241	0.0620	0.051	0.0177	0.69	0.68×10^6
200	0.241	0.0601	0.052	0.0182	0.69	0.58×10^6
250	0.242	0.0559	0.055	0.0192	0.68	0.42×10^6
300	0.242	0.0522	0.058	0.0204	0.68	0.31×10^6
350	0.243	0.0490	0.060	0.0216	0.68	0.23×10^6
400	0.245	0.0461	0.062	0.0227	0.67	0.18×10^6
450	0.246	0.0436	0.065	0.0239	0.67	0.14×10^6
500	0.247	0.0413	0.067	0.0250	0.66	0.11×10^6
600	0.250	0.0374	0.072	0.0271	0.66	0.070×10^6
700	0.253	0.0342	0.076	0.0291	0.66	0.044×10^6
800	0.257	0.0315	0.081	0.0312	0.66	0.033×10^6
900	0.260	0.0292	0.085	0.0338	0.65	0.024×10^6
1000	0.263	0.0272	0.089	0.0362	0.65	0.017×10^6
1200	0.269	0.0239	0.097	0.0402	0.65	
1400	0.274	0.0213	0.104	0.0442	0.65	
1600	0.278	0.0193	0.111	0.0471	0.65	
1800	0.282	0.0175	0.117	0.0512	0.65	
2000	0.286	0.0161	0.120	0.0534	0.65	

* ρ and a are for 29.92 in. Hg pressure.

Sources of data:

    c_p, μ, and k—Keenan, J. H., and J. Kaye, "Thermodynamic Proper-
        ties of Air," John Wiley & Sons, Inc., New York, 1945.

    ρ computed from simple gas law.

Notation: $a = g\beta\rho^2 c_p/\mu k$.

TABLE A-3. PROPERTIES OF GASES AND VAPORS

Gas or vapor	Chemical symbol	Boiling temp. at atmospheric press., °F	Density at atmospheric press. and at stated temp.		Specific heat at atmospheric press. and at stated temp.		
			°F	$\dfrac{lb_m}{cu\ ft}$	°F	c_p	c_v
Air.................	....	−318.0	32	0.0807	32	0.240	0.171
Ammonia..........	NH_3	−28.0	−28.0	0.0555	50	0.520	0.397
Carbon dioxide......	CO_2	−109.3	32	0.1227	70	0.205	0.160
Carbon monoxide...	CO	−313.6	32	0.0780	70	0.243	0.172
Freon-11..........	CCl_3F	74.7	74.7	0.3645	110	0.137	0.121
Freon-12..........	CCl_2F_2	−21.6	−21.6	0.3905	80	0.144	0.127
Freon-21..........	$CHCl_2F$	48.0	48.0	0.2852	60	0.136	0.116
Freon-22..........	$CHClF_2$	−41.4	−41.4	0.2950	120	0.157	0.133
Freon-113.........	$C_2Cl_3F_3$	117.6	117.6	0.4617	160	0.163	0.152
Freon-114.........	$C_2Cl_2F_4$	38.4	38.4	0.4889	110	0.163	0.150
Helium............	He	−452.0	32.0	0.0112	70	1.25	0.750
Hydrogen..........	H	−422.9	32.0	0.0056	70	3.42	2.44
Methane..........	CH_4	−258.9	32.0	0.0447	70	0.593	0.450
Methyl chloride.....	CH_3Cl	−10.8	−10.8	0.1591	70	0.198	0.159
Methylene chloride..	CH_2Cl_2	103.6	32.0	0.0236	70	0.154	0.131
Nitrogen..........	N_2	−320.4	32.0	0.0783	70	0.247	0.176
Oxygen...........	O_2	−297.3	32.0	0.0892	70	0.217	0.155
Sulfur dioxide.......	SO_2	14.0	14.0	0.1936	70	0.154	0.123

TABLE A-4. VISCOSITIES OF LIQUIDS AT 1 ATM*
(Coordinates for Fig. A-6a)

No.	Liquid	X	Y	No.	Liquid	X	Y
1	Acetaldehyde	15.2	4.8	56	Freon-22	17.2	4.7
2	Acetic acid, 100%	12.1	14.2	57	Freon-113	12.5	11.4
3	Acetic acid, 70%	9.5	17.0	58	Glycerol, 100%	2.0	30.0
4	Acetic anhydride	12.7	12.8	59	Glycerol, 50%	6.9	19.6
5	Acetone, 100%	14.5	7.2	60	Heptene	14.1	8.4
6	Acetone, 35%	7.9	15.0	61	Hexane	14.7	7.0
7	Allyl alcohol	10.2	14.3	62	Hydrochloric acid, 31.5%	13.0	16.6
8	Ammonia, 100%	12.6	2.0	63	Isobutyl alcohol	7.1	18.0
9	Ammonia, 26%	10.1	13.9	64	Isobutyric acid	12.2	14.4
10	Amyl acetate	11.8	12.5	65	Isopropyl alcohol	8.2	16.0
11	Amyl alcohol	7.5	18.4	66	Kerosene	10.2	16.9
12	Aniline	8.1	18.7	67	Linseed oil, raw	7.5	27.2
13	Anisole	12.3	13.5	68	Mercury	18.4	16.4
14	Arsenic trichloride	13.9	14.5	69	Methanol, 100%	12.4	10.5
15	Benzene	12.5	10.9	70	Methanol, 90%	12.3	11.8
16	Brine, CaCl₂, 25%	6.6	15.9	71	Methanol, 40%	7.8	15.5
17	Brine, NaCl, 25%	10.2	16.6	72	Methyl acetate	14.2	8.2
18	Bromine	14.2	13.2	73	Methyl chloride	15.0	3.8
19	Bromotoluene	20.0	15.9	74	Methyl ethyl ketone	13.9	8.6
20	Butyl acetate	12.3	11.0	75	Naphthalene	7.9	18.1
21	Butyl alcohol	8.6	17.2	76	Nitric acid, 95%	12.8	13.8
22	Butyric acid	12.1	15.3	77	Nitric acid, 60%	10.8	17.0
23	Carbon dioxide	11.6	0.3	78	Nitrobenzene	10.6	16.2
24	Carbon disulfide	16.1	7.5	79	Nitrotoluene	11.0	17.0
25	Carbon tetrachloride	12.7	13.1	80	Octane	13.7	10.0
26	Chlorobenzene	12.3	12.4	81	Octyl alcohol	6.6	21.1
27	Chloroform	14.4	10.2	82	Pentachloroethane	10.9	17.3
28	Chlorosulfonic acid	11.2	18.1	83	Pentane	14.9	5.2
29	Chlorotoluene, ortho	13.0	13.3	84	Phenol	6.9	20.8
30	Chlorotoluene, meta	13.3	12.5	85	Phosphorus tribromide	13.8	16.7
31	Chlorotoluene para	13.3	12.5	86	Phosphorus trichloride	16.2	10.9
32	Cresol, meta	2.5	20.8	87	Propionic acid	12.8	13.8
33	Cyclohexanol	2.9	24.3	88	Propyl alcohol	9.1	16.5
34	Dibromoethane	12.7	15.8	89	Propyl bromide	14.5	9.6
35	Dichloroethane	13.2	12.2	90	Propyl chloride	14.4	7.5
36	Dichloromethane	14.6	8.9	91	Propyl iodide	14.1	11.6
37	Diethyl oxalate	11.0	16.4	92	Sodium	16.4	13.9
38	Dimethyl oxalate	12.3	15.8	93	Sodium hydroxide, 50%	3.2	25.8
39	Diphenyl	12.0	18.3	94	Stannic chloride	13.5	12.8
40	Dipropyl oxalate	10.3	17.7	95	Sulfur dioxide	15.2	7.1
41	Ethyl acetate	13.7	9.1	96	Sulfuric acid, 110%	7.2	27.4
42	Ethyl alcohol, 100%	10.5	13.8	97	Sulfuric acid, 98%	7.0	24.8
43	Ethyl alcohol, 95%	9.8	14.3	98	Sulfuric acid, 60%	10.2	21.3
44	Ethyl alcohol, 40%	6.5	16.6	99	Sulfuryl chloride	15.2	12.4
45	Ethyl benzene	13.2	11.5	100	Tetrachloroethane	11.9	15.7
46	Ethyl bromide	14.5	8.1	101	Tetrachloroethylene	14.2	12.7
47	Ethyl chloride	14.8	6.0	102	Titanium tetrachloride	14.4	12.3
48	Ethyl ether	14.5	5.3	103	Toluene	13.7	10.4
49	Ethyl formate	14.2	8.4	104	Trichloroethylene	14.8	10.5
50	Ethyl iodide	14.7	10.3	105	Turpentine	11.5	14.9
51	Ethylene glycol	6.0	23.6	106	Vinyl acetate	14.0	8.8
52	Formic acid	10.7	15.8	107	Water	10.2	13.0
53	Freon-11	14.4	9.0	108	Xylene, ortho	13.5	12.1
54	Freon-12	16.8	5.6	109	Xylene, meta	13.9	10.6
55	Freon-21	15.7	7.5	110	Xylene, para	13.9	10.9

* By permission from "Chemical Engineers' Handbook," by J. H. Perry, McGraw-Hill Book Company, Inc., New York, 1950.

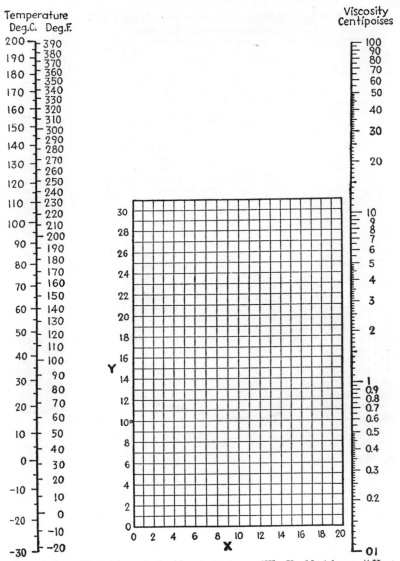

Temperature
Deg.C. Deg.F.

Viscosity
Centipoises

Fig. A-6a. Viscosities of liquids at 1 atm. (*W. H. McAdams, "Heat Transmission," McGraw-Hill Book Company, Inc., New York, 1954.*)

TABLE A-5. VISCOSITIES OF GASES AND VAPORS AT 1 ATM*
(Coordinates for Fig. A-6b)

No.	Gas	X	Y	No.	Gas	X	Y
1	Acetic acid	7.7	14.3	29	Freon-113	11.3	14.0
2	Acetone	8.9	13.0	30	Helium	10.9	20.5
3	Acetylene	9.8	14.9	31	Hexane	8.6	11.8
4	Air	11.0	20.0	32	Hydrogen	11.2	12.4
5	Ammonia	8.4	16.0	33	$3H_2 + 1N_2$	11.2	17.2
6	Argon	10.5	22.4	34	Hydrogen bromide	8.8	20.9
7	Benzene	8.5	13.2	35	Hydrogen chloride	8.8	18.7
8	Bromine	8.9	19.2	36	Hydrogen cyanide	9.8	14.9
9	Butene	9.2	13.7	37	Hydrogen iodide	9.0	21.3
10	Butylene	8.9	13.0	38	Hydrogen sulfide	8.6	18.0
11	Carbon dioxide	9.5	18.7	39	Iodine	9.0	18.4
12	Carbon disulfide	8.0	16.0	40	Mercury	5.3	22.9
13	Carbon monoxide	11.0	20.0	41	Methane	9.9	15.5
14	Chlorine	9.0	18.4	42	Methyl alcohol	8.5	15.6
15	Chloroform	8.9	15.7	43	Nitric oxide	10.9	20.5
16	Cyanogen	9.2	15.2	44	Nitrogen	10.6	20.0
17	Cyclohexane	9.2	12.0	45	Nitrosyl chloride	8.0	17.6
18	Ethane	9.1	14.5	46	Nitrous oxide	8.8	19.0
19	Ethyl acetate	8.5	13.2	47	Oxygen	11.0	21.3
20	Ethyl alcohol	9.2	14.2	48	Pentane	7.0	12.8
21	Ethyl chloride	8.5	15.6	49	Propane	9.7	12.9
22	Ethyl ether	8.9	13.0	50	Propyl alcohol	8.4	13.4
23	Ethylene	9.5	15.1	51	Propylene	9.0	13.8
24	Fluorine	7.3	23.8	52	Sulfur dioxide	9.6	17.0
25	Freon-11	10.6	15.1	53	Toluene	8.6	12.4
26	Freon-12	11.1	16.0	54	X, 3, 3-trimethylbutane	9.5	10.5
27	Freon-21	10.8	15.3	55	Water	8.0	16.0
28	Freon-22	10.1	17.0	56	Xenon	9.3	23.0

* By permission from "Chemical Engineers' Handbook," by J. H. Perry, McGraw-Hill Book Company, Inc., New York, 1950.

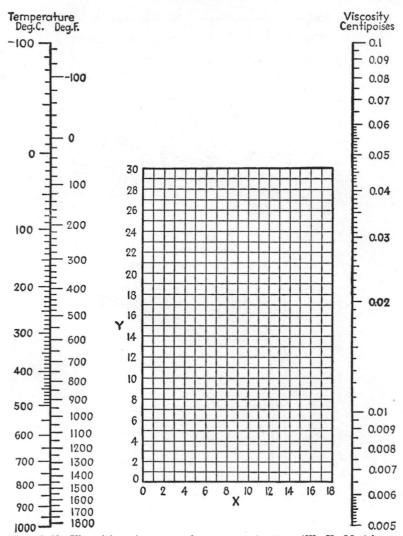

FIG. A-6b. Viscosities of gases and vapors at 1 atm. (*W. H. McAdams,* "*Heat Transmission,*" *McGraw-Hill Book Company, Inc., New York,* 1954.)

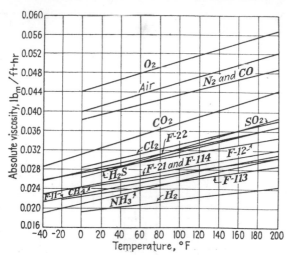

FIG. A-7. Viscosities of some commercial gases.

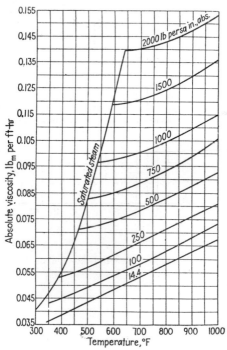

FIG. A-8. Absolute viscosity of superheated steam. (*ASME Fluid Meters Report*, 1937.)

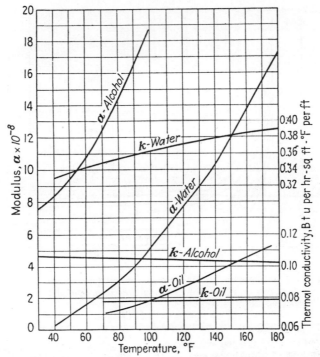

Fig. A-9. Values of the modulus, $a = g\beta p^2 c/\mu k$, and k for ethyl alcohol, water, and transformer oil. (*Values for alcohol and oil by W. J. King, Mech. Eng., May, 1932. Values for water from Table A-1.*)

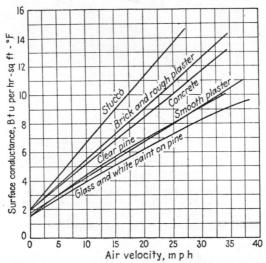

Fig. A-10. Surface conductances for various kinds of surface, for mean temperature of air and wall of 20°F. [*F. B. Rowley, A. B. Algren, and J. L. Blackshaw, Surface Conductance as Affected by Air Velocity Temperature and Character of Surface, Trans. ASHVE, 36 (1930).*]

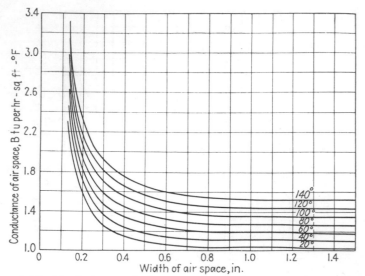

FIG. A-11. Conductance of air spaces for various mean temperatures. [F. B. Rowley, and A. B. Algren, Thermal Resistances of Air Spaces, Trans. ASHVE, **35**, 165–181 (1929).]

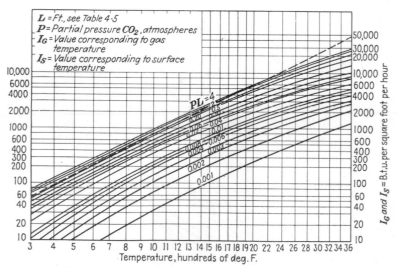

FIG. A-12. Total radiation due to carbon dioxide, to be used with Eq. (4-39). (W. C. McAdams, "Heat Transmission," McGraw-Hill Book Company, Inc., New York, 1954.) NOTE: For purpose of comparison the dashed line shows the radiation from a solid having a surface emissivity 0.10.

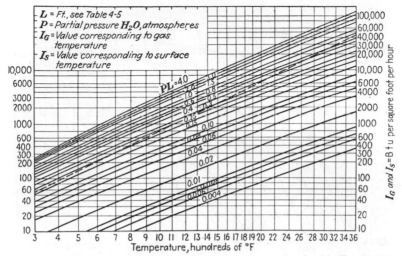

FIG. A-13. Total radiation due to water vapor, to be used with Eq. (4-39). (*W. C. McAdams, "Heat Transmission," McGraw-Hill Book Company, Inc., New York, 1954.*) NOTE: For purpose of comparison the dashed line shows the radiation from a solid having a surface emissivity of 0.10.

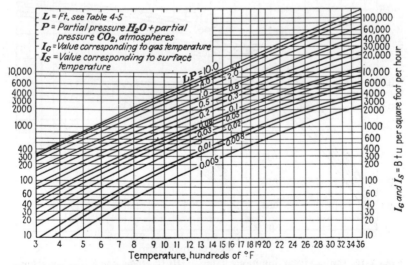

FIG. A-14. Total radiation due to water vapor and carbon dioxide, for a flue gas containing 0.8 mol of H_2O per mol of CO_2. (*W. C. McAdams, "Heat Transmission," McGraw-Hill Book Company, Inc., New York, 1954.*)

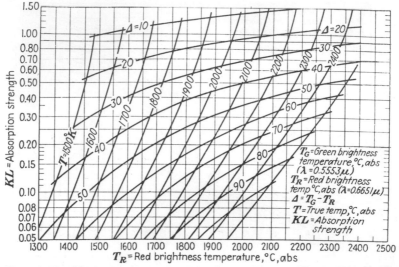

FIG. A-15. Absorptive strength of luminous flames. [*H. C. Hottel and F. P. Broughton, Ind. Eng. Chem., Anal. Ed.*, **4**, 166 (1932).]

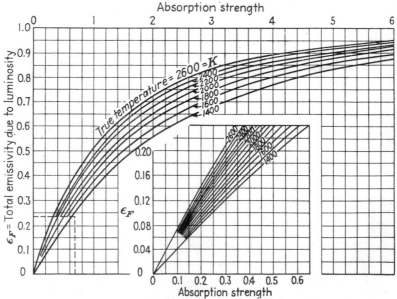

FIG. A-16. Emissivity of luminous flame. [*H. C. Hottel and F. P. Broughton, Ind. Eng. Chem., Anal. Ed.*, **4**, 166 (1932).] To determine the values of the green-brightness temperature T_g, the red-brightness temperature T_r, and the true temperature T in degrees Kelvin (i.e., degrees centigrade absolute), it is necessary to use an optical pyrometer, equipped with green and red lights, sighted at the flame. From these determinations the value of $\Delta = T_g - T_r$ is calculated, and the value of kL is then obtained from Fig. A-15. With this value of kL and the true temperature T, the value of pf is determined from Fig. A-16.

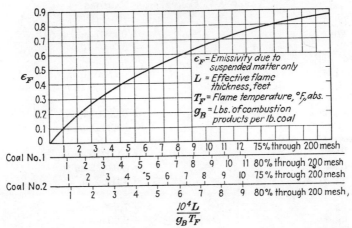

Fɪɢ. A-17. Radiation from powdered-coal flames. [*H. C. Hottel, Mech. Eng.*, **52**, 703 (1930).]

TABLE A-6. RADIATION FROM POWDERED-COAL FLAMES
[*From H. C. Hottel, Mech. Eng.*, **52**, 703 (1930).]

Coal number	Type	Btu/lb	V	ρ_o	ρ_1
1	Illinois bituminous, Saline-County . . .	12,800	38.4	84	75
2	McDowell, W.Va.	14,500	16.0	90	75

TABLE A-7. CONVERSION FACTORS

Length
- 1 cm = 0.3937 in.
- 1 m = 3.2808 ft

Area
- 1 sq cm = 0.1550 sq in.
- 1 sq m = 10.7639 sq ft

Volume
- 1 cu cm = 0.0610 cu in.
- 1 cu m = 35.3145 cu ft

Capacity
- 1 liter = 61.0250 cu in.
- 1 liter = 0.03532 cu ft
- 1 gal (U.S.) = 231 cu in.

Weight
- 1 lb_f = 7,000 grains
- 1 kg_f = 2.2046 lb_f

Pressure
- 1 kg_f/sq cm = 14.223 lb_f/sq in.
- 1 kg_f/sq m = 0.2048 lb_f/sq ft
- 1 atm = 14.696 lb_f/sq in.
- 1 atm = 1.0332 kg_f/sq cm
- 1 lb_f/sq in. = 2.3125 ft water at 70°F
- 1 in. water at 70°F = 5.20 lb_f/sq ft

Force
- 1 dyne $\times$ 10^6 = 1.020 kg_f
- 1 kg_f = 2.205 lb_f

Mass
- 1 slug = 32.17 lb_m
- 1 kg_m = 0.06854 slug

Gravity
- 1 ft/sec² = 12.96 $\times$ 10^6 ft/hr²
- 32.17 ft/sec² = 4.17 $\times$ 10^8 ft/hr²
- 1 ft/sec² = 30.48 cm/sec²
- 32.17 ft/sec² = 980.7 cm/sec²

Work, Energy, Heat
- 1 erg or dyne-cm = 10^{-7} watt-sec
- 1 joule = 1 watt-sec
- 1 g cal = 4.186 watt-sec
- 1 ft-lb = 1.356 watt-sec
- 1 Btu = 778.3 ft-lb
- 1 Btu = 1054.8 watt-sec
- 1 watt = 3.413 Btu/hr
- 1 kcal = 3.968 Btu

Heat capacity
- 1 joule/(g)(°C) = 0.23895 Btu/(lb)(°F)
- 1 g cal/(g)(°C) = 1 Btu/(lb)(°F)
- 1 Btu/(lb)(°F) = 1,900 watt-sec

TABLE A-7. CONVERSION FACTORS. (*Continued*)

Thermal conductivity

1 cal/(sec)(cm)(°C)	= 241.9 Btu/(hr)(sq ft)(°F/ft)
1 kcal/(hr)(m)(°C)	= 0.672 Btu/(hr)(sq ft)(°F/ft)
1 watt/(cm)(°C)	= 57.79 Btu/(hr)(sq ft)(°F/ft)

Surface conductance, Over-all coefficient

1 cal/(sec)(sq cm)(°C)	= 7373 Btu/(hr)(sq ft)(°F)
1 kcal/(hr)(sq m)(°C)	= 0.2048 Btu/(hr)(sq ft)(°F)

Viscosity

1 lb_m-hr/sq ft	= 12.96 × 10^6 slug/(ft)(hr)
1 slug/(ft)(hr)	= 32.17 lb_m/(ft)(hr)

See also page 84

INDEX